Award Winning

Science Fair Projects

Volume 1

By Sam Levine

Dedicated to Derric Lowery and Peter Collins

ACKNOWLEDGEMENTS

It was four years before the writing of this book, that one of my teachers first proposed that I may want to take a freshman high school course in which I would be able to do an independent science project. They were only going to accept a few of the 400 freshmen into the program, then known as "research methods physics," and before they would write my recommendation, they made sure I knew just how much work was going to be involved above and beyond the regular curriculum. I have to admit that it seemed like a daunting task at the time, but then I have always tended to take the road less traveled and always had fun along the way.

During my freshmen year, I had the opportunity to work under the supervision of two fantastic physics teachers. Karen Wolfson and Thomas Geyer worked with me to review my project ideas, point me in the right direction, and read multiple proposal drafts along the way. The project that resulted, an analysis of the effect of different variables on the output of a photovoltaic solar cell, ultimately was selected to compete at Worcester Regional Science and Engineering Fair, where it was then chosen to represent our region at the Massachusetts State Science and Engineering Fair (MSSEF). At the MSSEF, it won a second place award statewide, competing against over 200 of the best projects in Massachusetts.

As I prepared to enter sophomore year, and having experienced the excitement of science fair competition, I signed up for the Research Methods Chemistry and Mathematics course, where I was fortunate enough to be assigned to two of the best teachers I've ever met, Dr. Derric Lowery, and Peter Collins. Far beyond what the course itself involved, these instructors would really change my life, giving me more than just the knowledge I needed to succeed, but moreover the confidence in my abilities that would carry far beyond my work in their respective classes. That year, under their guidance, I designed the Near Space Experimentation Platform, which flew twenty miles up into the atmosphere as it traveled across the state. This project won first place at the state science fair, and was selected for an M.I.T. symposium selecting who would represent our state to compete internationally at the American Junior Academy of Science event in Vancouver. It all made for a very exciting year, and none of it would have been possible without the constant encouragement from Dr. Lowery and Mr. Collins. Additionally, I want to thank the head of our Science Department, Dave Hruskoci; my Principal, Todd Bazydlo; and our school Superintendent Dr. Joseph Sawyer, for consistently making these research courses available to students here in our town and for emphasizing the value of independent research.

Throughout my high school career, I've received fantastic support from many other individuals, who have always encouraged me to test my own limits, and push myself just one step farther. Susan Eriole, my guidance counselor, has been an invaluable advisor as I try to hone in on the many different options that I have to consider for my future. My very first guidance counselor, now my Assistant Principal, P.J. O'Connell, welcomed me to high school with open arms on day one, and has always been there for me to lean on, whether I needed advice, or just a friend to talk to.

Michelle Glidden, from the Society for Science and the Public contributed information on the background of the Intel International Science and Engineering Fair, and allowed me to share with you some of their materials on the rules and regulations behind the science fair program, and the judging process. Nancy Degon has been a leader in the science fair system in our region for many years. As chair of the Worcester Regional Science and Engineering Fair, she has been involved in evaluating hundreds of projects from the point of their original proposal, to the WRSEF event, and then their appearance at MSEF in the spring.

Naturally my parents have been very strong factors in my life, encouraging me to develop a love of learning, and allowing me to have access to a wide array of unique and amazing opportunities. Thanks to them, I've had the privilege of attending programs at M.I.T, Harvard, and W.P.I., where I was exposed to some unbelievable academic experiences, as well as wonderful professors and students alike. Through all the miles and miles of driving me around, the countless boxes of supplies arriving at our door, and even the turning of our finished basement into a carbon fiber lamination facility, they've always been nothing less than 100% supportive of everything I do.

Last but not least, I must thank my fellow competitors from the 2011 Massachusetts Science and Engineering Fair, who contributed their projects for this book. These students were incredibly generous in their willingness to share their knowledge and experience with other young scientists from around the nation. The primary goal of this work is, after all, to provide a guideline for other students considering a project of their own. By looking at the wide variety of different projects in different fields, my hope is that others can glean first hand experience in just how a winning project is put together from conception to competition.

Foreword

On September, 12 1962, President John F. Kennedy stated that, "We choose to go to the moon in this decade and do the other things, not because they are easy, but because they are hard. Because that goal will serve to organize and measure the best of our energies and skills, because that challenge is one that we are willing to accept...". This statement fueled the dreams of school children and adults throughout America and initiated one of the greatest science projects in human history. For many of the young scientists and engineers who would work on projects directed at this effort, this would be their first real venture into long-term, focused scientific investigations. Many had not experienced research methods and investigation techniques in high school or even at the university level unless they had completed graduate work in their field.

In the decades since the challenge to go to the moon, K-12 science educators have placed a greater emphasis on science projects directed at building skills in the methods of research. In many areas of the country, school, regional and state science fairs allow students to communicate their work to others and to compete with other students in similar categories. However, these opportunities aren't universally available and while some educators and students would like to participate, they are not sure how to get started. Even with the opportunity, one of the greatest challenges for a student is deciding on a topic and having a feel for the science fair process.

Within this work, Sam Levine has assembled a guide to help both students and educators navigate the science fair process from idea conception, to the myriad of required forms and approvals, to the final presentation. Having participated in a number of science fairs and winning first place in his category at the 2011 Massachusetts State Science and Engineering Fair (MSSEF), Sam is an authority in the process and has gleaned a great deal of knowledge that he willingly shares with readers who want to know more about the opportunities and processes required for a successful project.

As mentioned previously, one of questions that students always ask at the outset is, "what topics should I consider"? Perhaps one of the most valuable aspects of this work is the collection of abstracts contributed by many of the participants in the 2011

MSSEF. These abstracts offer a glimpse into the broad expanse of topics that may be investigated as part of an independent high school research project.

With this work, Sam has given a gift to students, educators and administrators that I am sure will be a valuable asset as students pursue their independent research projects. These projects will introduce them to the skills and thinking processes that will serve them well in their future educational and career endeavors.

Derric Lowery, PhD

Welcome Science Fair Friends and Fans

For years, science fairs have existed to retrieve the best and brightest students in the world from their classes and introduce them to the scientific community, all the while revealing brilliant ideas of students', many of whom turn into quite intrepid inventors.

Students from countries all over the world compete in a number of fields from Biology and Chemistry to Engineering and Ecology and everything in between, with projects of various derivations. The one thing each project has in common is that the student, or team of students, hopes to rank among the best in their country, state, or province to reach that top-tier fair, the International Science and Engineering Fair, where they will compete for scholarships, cash prizes, patent sponsorships, and internships from a number of top-notch companies in their field of study.

There are a number of ways to enter the science fair, but the simplest is to start at your middle school or school fair. If your school does not have a science fair program, you should check your regional fair's website, or the state science fair's website.

Working your way to national and international recognition passes you through many steps, which often begin in your classroom, and can wind up in Washington DC San Diego,

Vancouver, BC or a host of other cities. The achievement is incredibly rewarding, both intellectually, and even financially, and testing your abilities against others can be a truly life altering event for a student who likes a challenge.

In the United States, the basic hierarchy of science fairs is based on the general process, which typically starts at your school. *

Working with a teacher, mentor volunteer, or other science professionals, you will conceptualize a project for your school science fair, prepare a proposal, develop an experimental design, and carry out weeks or months of experimentation and data analysis which will culminate with your presentation at your local school science fair.

Once the projects are judged at your school, usually by teachers or other parent volunteers, the organizers will select the top 10-20 to move on to the Regional Science Fair level, depending of course on the size of your school.

When you compete at the Regional Science Fair, against the top projects from schools in your area, you'll be immediately amazed at how different the caliper of work is. Suddenly, instead of a superstar, you will quickly find that you are just one of many great projects, all of them well researched, meticulously executed and beautifully presented. After a full day of judging, projects will be selected for the grand prize, as well as first, second and third place awards, as well as a host of others. Most of them carry with them a financial award that can range anywhere from one hundred to three thousand dollars, all of it donated by very generous sponsors. The best of these projects at each of the regional science fairs, usually those no lower than third place, are selected to compete in the State Science Fair, usually held in May.

If you've made it this far, and have the honor of being invited to compete at the state level, you will see projects that are nothing short of mind blowing. No more are you looking at a study of the reaction time of a goldfish to a bright light, or a chart of battery life in flashlights. Here at the state level, it's a whole different world. As you set up your display, and then walk around to see your fellow students work, you'll find people who worked to study the effect of different chemotherapeutic agents on cancer cells, a fuel cell using bacteria as a power source, or a solar salt water distillation system. On the other side of the room, you'll find a student who designed a unique iPod application, one who harvested seaweed to make biodegradable bags, or a sub-orbital space platform that conducts high altitude experimental research before being tracked as it descends to earth a hundred miles away. Yes, these projects are "the real deal," and the students you meet here will likely go on to great things down the road. Many have competed at the state level multiple times, so be prepared to be at your best. Looking sharp, speaking clearly, and following the advice of those who have been there before, will take you a long way.

At this level, the expectations, and certainly the competition are much higher than you will have seen. The judges are often selected as experts in their field, and specifically assigned to projects with which they will have quite a bit of scientific familiarity. At the end of a day or two of anxious presentations, the award recipients are announced. These prizes can go as high as a full four-year scholarship, worth $40,000 or more. Most first place awards are in the $1500-$5000 level, and others include an opportunity for coveted internships with the sponsor companies.

The final step is for the state science fair body to select delegates to attend the international science fair. Here in Massachusetts, that is done by inviting all of the First Place winners from the Massachusetts Science and Engineering Fair to come back to the Boston area in October, and do a short presentation before a team of scientists. The judges will then select a handful of students to attend the International event in February. Of course your state will likely have it's own unique method of determining these delegates.

Merely entering the science fair is a great privilege and it is an amazing opportunity for any student who wishes to participate. I have competed for the last two years, and I can say that it is the vast array of student entries that make the science fair one of the best childhood experiences in my recent memory. All of the hard work students do to put forth their best foot in front of their local scientific community is recognized in this format throughout the world, and it is something that I believe is partially responsible for many children's interests, including myself, in various scientific fields. To not only "enter" the fair, but to achieve success, is something that can easily change the direction of a young students life.

There are a number of critical components that make up a successful science fair project; the main ones are presentation/knowledge of subject, complexity, thoroughness of research, precision of experimentation and data, accuracy of deductions, and practicality of subject.

Presentation is paramount in pitching anything in any setting; science fairs are no exception. It all begins with choosing a project that you know a lot about and doing further research to the point where you can be questioned by a professional in the field and respond to their inquiries correctly. At top-tier science fairs, this is especially important, because you do not want to seem uneducated about your physics project when being judged by the Ph.D. physicist. Similarly, you do not want to over-learn your topic to the point where you mention irrelevant things, because this will only detract from the significance of the work you have done.

Complexity is often well rewarded by science fair judges

who recognize just how much work the student scientist has put into the project. Complicated projects alone will not win a science fair; good presentation of a 10-minute project will not win a science fair; combining, and maximizing, complexity and knowledge of the subject, however, will have a synergistic effect that will do wonders with the judges. The goal of the project should be to go in depth enough to create a comprehensive, thorough project that accurately depicts just how much you know about the topic, and allows you to conduct independent research. If you do not know much about the topic and do not want to choose another, you are in luck, because almost every science fair judges rubric has a research grade included. Do not hesitate to research your topic, as it can only help you perform better at the science fair and aid in conducting your experiment, as well as become a better researcher down the road.

Precision is not paramount to every science fair project; only to the successful ones! Individual experiments can only have one responding/dependent variable, and you will need to point out in your report and poster, which variable is which. Obviously, your data must be accurate. It is for that reason that you must always show your experiment to someone else with a highly respected scientific opinion, just as one would have an essay proofread by a good writer. Sometimes after we look at things for a long time mistakes fade into the background, but this method should act as a failsafe for that. Data must also be precise. Be absolutely sure that you always use the proper significant figures and mention your margin of error (sometimes error bars are necessary on graphs). One common mistake that can cost points on a judging is the failure to use the metric system for all measurements. You didn't use a ¼" screw to hold on the cover, but rather a 6.35mm screw.

Having done a project with adequate complexity and performed precise experiments, you now need to make sure to make the correct deductions from your data. When inaccurate deductions meet lack of precision, odd anomalies can occur, as in the story describing the idea of the flawed theory of spontaneous generation of organisms.

Spontaneous generation is the idea that nonliving things can become living things. This theory survived for a long time as it was supported by a number of faulty experiments. Many, many experiments refuted the theory, but one experiment kept it afloat. A scientist by the name of John Needham performed an experiment where he heated a beaker of gravy to kill the bacteria inside and then sealed it. After this, he observed it for days and soon it was teeming with bacteria. As a believer in spontaneous generation, Needham deduced that the bacteria had come from the gravy. In actuality, following experiments led to the conclusion that Needham had not heated the beakers to boiling, which would have successfully killed the bacteria. While his experiments were logical, he should have tried different temperature and boiling times, which would have ultimately killed the endospores of the bacteria. Instead, he saw what he wanted to see, and deduced what he wanted to deduce.

Do not let this be you; if your hypothesis is proven wrong, that's fine. It is better to analyze it correctly. It actually reflects very well on you to admit that you were wrong, and you can always draw upon the ideal that you learn more from being wrong than from being right. In my very first science project, my precise research completely disproved my original hypothesis, yet I still achieved a second place award for the state of Massachusetts. Don't be afraid to go back and do further research, making plans for future experiments. Showing a thirst for the answer is only going to make you look brighter and more inquisitive to judges. Just remember, being wrong does not make you any less bright, only walking away from a great scientific investigation you have started can do that.

Practicality is another key element that judges are looking for. It is true that Da Vinci was the first to think of how a "flying machine" (helicopter) would look and function and what we call a tank today. However, he did not have the resources to make either. Although those ideas were perfectly feasible, he was in the wrong time to make them realities. The same principles must apply to science fair projects; you should not get in over your head on a project that is too complex to manage. You should do as complicated and involved work as you can, but not to the point where you do not have the resources to make your project work. In order to expand your horizons, you can work with an existing laboratory, but do not try to invent the next flying machine or tank without sufficient resources available.

Entering the science fair is a great privilege and it is an opportunity for any student who wishes to enter. It is the vast array of student entries that make the science fair one of the best childhood experiences in my recent memory. All of the hard work students do to put forth their best foot in front of their local scientific community is recognized in this format throughout the world, and it is something that I believe is partially responsible for many children's interests, including myself, in various scientific fields.

Included in this book is a collection of science project proposals that I collected from a number of very generous students at the last MSSEF (Massachusetts State Science and Engineering Fair), where I competed in 2010. These students stepped up and donated their projects to this work, solely in the hope of helping other young scientists down the road, and I'm extremely grateful to them all. They range from simple to extremely complex projects, and are presented here not as abstracts, but as full final project reports. The objective of this book is not for students to attempt to copy the work of others, reproduce their experiments, or even use them as a starting point for other research. Rather my hope was merely for you to see the many different styles used, and to see how these top students took a project from conception to competition. By looking through a host of award winning projects, the hope is that it will provide the motivation, guidance, and direction you need to make your science fair dreams a reality.

* Note that this information on the science fair process is based on the Intel Science and Engineering Fair, ISEF, which is one of the largest such program in the world. There are many other fairs, each with their own rules and regulations, such as the Siemens Competition, Discovery Young Scientist Challenge, Canada Wide Science Fair, and Intel Science Talent Search.

About the Intel® Science and Engineering Fair

The Intel® International Science and Engineering Fair® (Intel ISEF), the world's largest international pre-college science competition, annually provides a forum for more than 1,500 high school students from over 65 countries, regions, and territories to showcase their independent research.

The fair began in 1950 as the National Science Fair created by Society for Science & the Public (then known as Science Service). At this initial fair, there were 30 finalists representing 13 fairs within the United States. In 1958, the fair became international for the first time when Japan, Canada, and Germany joined the competition. Over the decades, the fair has continued to grow, reaching over 500 finalists in 1982 and over 1000 finalists in 1995. Intel became the title sponsor of the ISEF in 1997.

Today, millions of students worldwide compete each year in local and school-sponsored science fairs; the winners of these events go on to participate in the approximately 500 Intel ISEF-affiliated regional, state and national fairs around the world from which the best win the opportunity to attend the Intel ISEF. The Intel ISEF unites these top young scientific minds, showcasing their talent on an international stage, enabling them to submit their work to judging by doctoral level scientists—and providing the opportunity to compete for over $4 million in prizes and awards.

Society for Science & the Public provides online resources for students interested in conducting independent research as well as for the parents, teachers and fair directors guiding these efforts. Notably, it is recommended that everyone review the International Rules for Pre-college Science Research: Guidelines for Science and Engineering Fairs. In addition to providing the rules of competition, these rules and guidelines for conducting research were developed to:

- protect the rights and welfare of the student researcher and human participants,
- protect the health and well-being of vertebrate animal subjects,
- follow federal regulations governing research,
- offer guidance to affiliated fairs,
- use safe laboratory practices,
- address environmental concerns.

Please visit **www.societyforscience.org/isef** for the most current information and resources to aid in the successful completion of an independent research project.

Michele Glidden
Director of Science Education Programs
Society for Science and the Public

Guidelines for Projects

As you begin to select the topic for your project, you may be inclined to do research that involves the use of human or animal subjects. Before you even begin to go down that road, you should familiarize yourself with the specific guidelines that these types of projects require.

The following information is provided by the Society for Science and the Public as a guide for all researchers involving the risk assessments for human and animal experiments. In addition to the information below, a full Rules and Regulations section for the Intel Science and Engineering Fair can be found at this link http://www.societyforscience.org/document.doc?id=311. Other science fairs may have their own unique guidelines, and you are encouraged to read these thoroughly, long before you being your project.

"The purpose of this guide is to assist student researchers, teachers/mentors and local IRB's as they evaluate risks and design research projects that respect the rights and welfare of human subjects. The complete Human Subjects rules and guidelines can be found on the Web at: www.societyforscience.org/isef/document/.

This document contains information on the following topics:
1. Risk Assessment and Risk Reduction
2. Types of Risks and Suggestions for Reducing Risk

A. Risk Assessment and Risk Reduction

Risk Assessment involves consideration of **physical** and

psychological risks along with the **protection of privacy.** The student researcher, adult sponsor and qualified scientist must develop procedures that reduce and minimize any risks to human subject participants.

The IRB will review the Research Plan and make the following determinations:

- whether the study is approved or must be revised
- whether the study contains no more than minimal risk or more than minimal risk to potential participants. The IRB will consider characteristics (e.g., age, health status, vulnerability to coercion) of the study population, the specific risks (e.g., physical, psychological, social, privacy) associated with the research activity and local norms when making a risk level determination;
- whether documentation of informed consent/subject assent and/or parental permission are required or can be waived
- whether a qualified scientist is required

No more than minimal risk exists when the probability and magnitude of harm or discomfort anticipated in the research are not greater than those ordinarily encountered in daily life or during performance of routine physical or psychological examinations or tests. Studies must involve **anonymous data** to be considered no more than minimal risk.

More than minimal risk exists when the possibility of

physical or psychological harm or harm related to breach of confidentiality or invasion of privacy is greater than what is typically encountered in everyday life.

B. Types of Risk
1) Physical Risks:

a. **Exercise** other than ordinarily encountered in daily life *by that subject* would be considered more than minimal risk. One must consider characteristics of potential research subjects as well as the type of exercise involved in the study.

Examples:
- Walking the length of standard hallway

For most healthy subjects, this activity could be considered "minimal risk." For the elderly or someone recovering from knee surgery, this might be considered "more than minimal risk."

- Swimming 500 meters

For the general population, this activity would be considered "more than minimal risk." For members of the varsity high school swim team, this activity could be considered "minimal risk."

b. **Ingestion, tasting, smelling, application of a substance** that pose any health risk are considered "more than minimal risk". Ingestion or tasting projects that involve commonly available food or drink will be evaluated by the IRB who will determine risk level based upon the nature of study and local norms around food typically encountered in the research setting. For Example:

- Some school IRBs may consider a tasting study minimal risk based on the fact that the food being studied is commonly available to all students in their school.
- Conversely, an IRB at another school may deem the same study more than minimal risk if the food being studied is not commonly available to students or they believe that parents in their community would want to provide parental permission before their minor child could participate in the study.

c. **Medical examples**

Blood glucose testing with a glucometer that is conducted by a diabetic on a daily basis could be considered minimal risk. However, it would be considered more than minimal risk when a glucometer is used by a subject who does not perform this test as a function of their daily life. Student researchers must receive training by a qualified scientist on the proper technique of capillary blood glucose sampling. Risks to the subject include pain, infection, and injury and risks to the researcher could include possible exposure to blood/body fluids, or needle stick.

A project involving the measurement of blood pressure in which a student researcher uses a commercially available automatic blood pressure device would be considered a minimal risk study. The study would be considered more than minimal risk if a manual sphygmomanometer were used. Risks include vascular spasm, nerve damage, and bruising due to improper technique. The IRB must examine the context of the research plan to ascertain these risks. Training of the student researcher should be required by the IRB. The IRB may also require a qualified scientist. Most often, these measurements are obtained in conjunction with exercise. If that is the case, the IRB should refer to item 1a above to assess the overall risk to the subject. Each research plan that employs vital sign measurements should also include a plan of how to deal with vital signs measurements that are out of range. For example, when a reading is obtained that is outside of the normal range, the person should be referred to their healthcare provider or the nearest emergency facility.

2) Psychological Risks

A research activity (e.g. survey, questionnaire, viewing of stimuli) or experimental condition that could potentially result in **emotional stress** would be considered **more than minimal risk.** For example, answering questions related to personal experiences such as sexual or physical abuse, divorce and/or psychological well being (e.g. depression, anxiety, suicide) must be considered more than minimal risk and should have documented informed consent/minor assent/parental permission (as applicable).

Additionally, research activities that involve exposing subjects to stimuli or experimental conditions that could potentially result in emotional stress must also be considered more than minimal risk. Examples include violent or distressing video images, distressing questions, materials or activities that could potentially result in feelings of depression, anxiety, or low self-esteem in subjects.
Reducing Risk associated with Emotional Distress: Care must be taken to try to reduce potential emotional distress. For example, to reduce risk in a study involving a survey about depression and suicide, consider having a school counselor available to talk with students if they are feeling distressed or having a statement at the end of the survey directing students to the school counselor or school psychologist.

3) Risks due to Invasion of Privacy & Breach of Confidentiality

The student researcher and the IRB must consider whether any activity could potentially result in negative consequences for the subject due to **invasion of privacy or breach of confidentiality.** For example, if the study involved collecting a student's GPA and the data were accidentally made available to unauthorized persons, the research subject could suffer embarrassment and feelings of distress related to the invasion of his privacy.

Reducing Risk:

Risk level can be reduced by appropriately protecting confidentiality or collecting data that is anonymous and uses data collection procedures that make it impossible to link any identifying information with his/her responses or data.

a) **Anonymity** involves collecting research data in such a way that it is impossible to connect research data (e.g. responses, questionnaires) with the individual who provided the data. That is, personal identifiers (e.g. names, birthdates, social security numbers) are not collected. **Whenever possible, student researchers should collect data anonymously.** (While collecting data anonymously does reduce risk, not all anonymous studies are considered minimal risk.)

- To collect data anonymously, student researchers must not require subjects to give their name or any other identifiable information (birth date, email address, etc.)
- If documented informed consent, assent, and/or parental permission is/are required, the forms must always be kept in a secure location separate from the data.

b) **Confidentiality** is necessary when personal identifiers such as name, birth date, telephone number, photograph, email address or mailing/street address are collected.

- Protecting confidentiality involves taking careful measures to ensure that the research data and/or responses are not disclosed to the public or unauthorized individuals with identifiable information. When research activities involve collection of personal information (e.g. history of abuse, drug use, opinions, fingerprints, emotional functioning, grades) or health-related data (genetic material, blood, tissue) the researcher must consider risks related to invasion of privacy and possible breach of confidentiality.

- If the research involves data from the same subject on multiple occurrences, the data or survey would need to be labeled with an identifier to be linked with the data collected at a later date. In this case, confidentiality could be maintained by labeling the surveys or data with a subject number and keeping a list of names and subject numbers in a separate and secure (e.g., locked file cabinet, password protected computer) location. Once the 2nd round of data is collected, the surveys/data may be matched using the subject number and any identifiers should be removed from the data/surveys. At this point, the list of names and subject numbers should be securely discarded (e.g., shred). If documented informed consent, assent, and/or parental permission is/are required, the forms must be kept in a secure location separate from the data.

Special Considerations:
Threats to Anonymity

- If the number of participants is relatively small and/or all participants are from an identifiable source (e.g., an English class, softball team), the anonymity of the data could be threatened. That is the student researcher or anyone with access to the data could potentially link the survey responses to an individual. In addition, presenting the results of the study (even in aggregate) could threaten the subjects' privacy or result in negative consequences for the subjects.
- If informed consent/assent/parental permission forms (which include names) are collected and the sample is relatively small, it could be possible for the student researcher or an unauthorized person to link the survey responses with subjects.

Making Data Anonymous

• Sometimes data may not be collected anonymously, but can be made anonymous after data collection. For example, if the student researcher uses interviews or observations to collect the data, the data would not be anonymous at the time of collection. However, if names are not collected or are removed from the data soon after collection, the data set would then be anonymous.

Risks Related to Threats to Anonymity

- Be sure to consider any ramifications of the student researcher being able to link responses with subjects. Most importantly, would there be any negative consequences for the research subjects if the student researcher could link responses with the subjects. This is especially important when the research subjects are peers to the researcher. When the subjects are peers of the student researcher, the researcher/QS/IRB should give extra consideration to any potential risks related to the student researcher having knowledge of his/her peers' data (e.g., grades, body weight, etc). To eliminate such risks, it may be prudent to have an adult collect the data and hand it over to the student research after identifiers are removed and it is anonymous.

- Be sure to consider the possibility of and ramifications of an unauthorized person (e.g., another student, parent, teacher, administrator) getting access to the data and being able to link responses to individual subjects or groups of subjects (e.g., softball team).
- Consider the nature of the study/data collected. Issues of anonymity and confidentiality are most salient for studies involving sensitive and personal information. Examples of data that should receive special consideration include grades, health/mental health information, experiences of child abuse, illegal behavior, socially unaccepted behavior, anything that could cause the subject embarrassment or legal or disciplinary negative consequences.

4) Risk Groups:

As noted above, the physical, psychological and other risks

of participation in a study may depend on the specific sample of subjects involved. The physical risk of an activity such as jumping roping will be much higher for an elderly (or even middle aged subject) than for a middle or high school subject. In contrast, the risks of a breach of confidentiality or anonymity would be greater for a group of high school students answering questions about alcohol use than for a group of older adults for whom it would be easier to collect the data in a anonymous fashion.

Some groups deserve special consideration. If the research study includes subjects from any of the following groups described below, the student researcher and the IRB must consider whether the nature of the study requires special protections or accommodations for subjects in these risk groups.

1) Any member of a group that is naturally at-risk (e.g., pregnant women, mentally disabled persons, economically or educationally disadvantaged persons, individuals with diseases such as cancer, asthma, diabetes, cardiac disorders, psychiatric disorders, learning disorders, etc.). The nature of the study is an important consideration when determining if special protections are required. For example, special protections would not typically be necessary to include pregnant women in a study involving performance on a cognitive test or completion of a simple survey.

2) Special vulnerable groups that are covered by federal regulations (e.g. children/minors, prisoners, pregnant women,

students receiving services under the Individuals with Disabilities Education Act). Specifically, the IRB and the student researcher should consider whether potential study participants who are receiving services under the Individual Disabilities Education Act need special accommodations and/or are appropriate for inclusion in the study as a research subjects. For example, an IRB may choose to require parental permission for minor subjects receiving special education services even when parental permission has been waived for general education students. Confidentiality must be maintained so as not to identify/isolate students.

C. **InformedConsent**

Informed consent refers to the process of ensuring that potential human subjects understand that they may choose whether or not to participate in a study. Individuals should never be forced or coerced to participate in a research study. A teacher, school administrator or anyone requiring students to participate in a research study as a human subject would be considered a serious violation of informed consent principles. That is, the research subject must freely decide to participate and not feel coerced or forced into doing so.

To make an informed decision about whether an individual wants to participate, the human subjects must be informed about what they will be asked to do and if there are any risks or benefits involved. For example, if the subject will be asked to complete an interview or a survey, the nature of the survey should be described (e.g., questions about emotional functioning, students experiences around divorce, grades and

SAT scores). In most cases, the informed consent process also includes a description of the purpose of the study. However, in rare circumstances detailed information about the purpose of the study will not be included if purpose of study requires innocuous deception that poses minimal risk to the human subject. A school's IRB may allow innocuous deception studies such as a study designed to determine if colored paper affects the time it takes a student to complete a given written task. The IRB may require a QS to help develop appropriate informed consent procedures which respect the rights of human subjects but do not threaten the validity of the study. Subjects 18 years and older **must** be provided with all of the information mentioned above and give their **Informed Consent** before participating in a research study. In most cases, if subjects are under the age of 18, a parent or legal guardian must be presented with all of the information described above before giving **Parental Permission** for their minor child to participate. **Minor Assent** refers to procedures giving developmentally appropriate information to children and to adolescents about the study and giving them a choice as to whether or not they will participate. High school students should be supplied with ALL of the information mentioned above and give their verbal and/or written assent to participate.

Obtaining Written Informed Consent, Parental Permission or Minor Assent

An informed consent form is typically used to provide written information to the human subject or parent/guardian and to document written informed consent/parental permission/minor assent. This form typically includes the purpose of the study, what the subject will be asked to do, the nature of any surveys, questionnaires or interviews, any risks and any benefits to the subject. The form should also contain information that explains to the potential research subject or parent/guardian that participation in the study is voluntary and that the subject is free to stop participating at any time. The Sample Informed Consent Form provides an example of how this information can be presented.

A copy of any survey or questionnaire must be attached to the form when parents/guardians are being asked to give their permission for their minor child to participate. This process allows the parent to review the material to which their child will be exposed and make an informed decision about whether they want their child to participate. However, in some cases sending home a copyrighted survey may be a violation of the test publisher's regulations. In other cases, sending home a copy of the survey may threaten the validity of the study. The IRB must decide whether an appropriate description of the survey on the Sample Informed Consent Form provides enough details so that the parent or guardian can make an informed decision.

Waiver of Written Informed Consent/Parental Permission/Minor Assent

Obtaining informed consent from an adult or minor assent is always required. However, the IRB may waive the requirement for documentation of written informed consent/assent and/or parental permission if the research involves **only minimal risk and anonymous data collection and if it is one of the following:**

a) Research involving normal educational practices

b) Research on individual or group behavior or characteristics of individuals where the researcher does not manipulate the subjects' behavior and the study does not involve more than minimal risk. c) Surveys and questionnaires that are determined by the IRB to involve perception, cognition, or game theory and do NOT involve gathering personal information, invasion of privacy or potential for emotional distress.

d) Studies involving physical activity where the IRB determines that no more than minimal risk exists and where the probability and magnitude of harm or discomfort anticipated in the research are not greater than those ordinarily encountered in DAILY LIFE or during performance of routine physical activities.

As explained above, informed consent/minor assent or parental permission is always required. It is merely the process of obtaining a signature to document informed consent/minor assent or parental permission that can be waived in the circumstances mentioned above. If there is any uncertainty regarding the appropriateness of waiving written informed consent/assent or parental permission, it is strongly recommended that documentation of written informed consent/parental permission be obtained. In addition, it is recommended that parental permission not be waived for minor participants who are younger than high school age.

D. Online Studies

Studies that collect data via use of the internet (e.g., email, web based surveys) require special consideration from the IRB. The use of the internet for data collection will pose challenges in

a) collecting anonymous data (e.g., IP addresses are recorded by many online survey tools), b) obtaining informed consent and

c) ensuring that participants are of the appropriate age to give informed consent.

The research plan must explicitly address how these challenges were evaluated and addressed.

Guidelines to school IRBs:

1. 1) It is recommended that studies deemed to be "more than minimal risk" (e.g., about

personal/sensitive issues) only be conducted online if the student is working with a qualified scientist who has experience conducting IRB approved research at a Research Institution, University or College. Because IP addresses are gathered by many online survey tools, specialized procedures are needed to ensure that the data is collected anonymously. Ideally, this type of research should be done through a Research Institution, University or College with formal IRB approval by the institution.

2. 2) Studies deemed to be "minimal risk" and targeted to adult subjects (18 years and older) can be conducted online and/or subjects can be recruited by email. The research plan must address how the researcher will ensure that only adult subjects are actually recruited. See below for more information about what is needed in the informed consent process for an online survey.

3. 3) Studies deemed to be "minimal risk" that include minor subjects (under the age of 18 years) can be conducted online with the subjects' parental/guardian permission. In this case, the parent/legal guardian must give consent using a traditional, paper consent form before the minor participant completes the online survey. After formal parental permission is secured, the researcher can email or give the subject a link to the online study. As always, minor assent procedures must also be included. That is, the minor must be given the same information given to the parent/guardian and must be aware that participation is completely voluntary. See below regarding what information must be given to parents/guardians and minor subjects to secure parental permission and minor assent.

4. 4) It is strongly recommended that members of the IRB test the link the student will provide to human subjects to complete the online study to ensure that

the correct procedures for obtaining adult consent and/or minor assent are in place.
5. 5) The student researcher, adult sponsor or QS, and at least one member of the IRB should be knowledgeable about the specific online survey tool (e.g., whether IP addresses or other identifiers are gathered, how secure the online survey is, who has access to responses, whether responses are able to be securely deleted, etc).

Procedures for Obtaining Adult Subjects' Consent and Minor Subjects Assent for Online Surveys

1) All information required for the Informed Consent/Assent Process must be presented to the potential research subject before the survey begins.

2) The following statement or something similar must also be included:

There is always the possibility of tampering from an outside source when using the internet for collecting information. While the confidentiality of your responses will be protected once the data are downloaded from the internet, there is always a possibility

of hacking or other security breaches that could threaten the confidentiality of your responses. Please know that you are free to decide not to answer any question.

3) The survey should be set up in a way that the potential subject must click on a "button" or type in a response indicating that he/she has read the consent/assent information

and agrees to participate to take the potential subject to the actual survey. That is, the survey questions are not viewed by subject until he/she clicks on or types in a response to indicate his/her voluntary participation.

Extra Information to be Included on the Parental Permission Form

1) Parental permission cannot be obtained online. When required, a traditional, paper form must be signed by the parent before a minor can participate.

2) In addition to all of the information required for parental permission, the following statement or a similar statement must also be included:

There is always the possibility of tampering from an outside source when using the internet for collecting information. While the confidentiality of your child's responses will be protected once the data are downloaded from the internet, there is always a possibility of hacking or other security breaches that could threaten the confidentiality of your child's responses. Your child will be instructed that he/she is free to decide not to answer any question.

Procedures for Protecting Confidentiality Related to Downloading of Data

- If IP addresses are collected by the survey tool, the addresses should be deleted from the downloaded data file. All responses should then be deleted from the online survey. The resulting data file that is used for data analysis should be free of any identifier, including an IP addresses or other electronic identifiers.

- The data file should be stored on a password protected computer. Any back up data files should be stored in a secure location.

Examples of Research Studies with Acceptable Suggested IRB Decisions note that IRB's have the prerogative to be more conservative.

- Student wants to compare the amount of television and type of television shows viewed by boys and girls.

Minimal risk study: Parental permission not required if data are collected anonymously and if subjects informed of voluntary nature and right to withdraw at any time.

- Student researcher wants to examine the relationship between favorite restaurant and weight in 9 – 12 graders.

More than minimal risk study: Parental permission required because of emotional risks and impact on self esteem associated with a student reporting on his/her weight. Even with parental permission, procedures for anonymous data collection should be used. Care should be taken to ensure that the student researcher is not able to link data with a particular subject.

- Correlate television viewing with mood

Potentially more than minimal risk study: Parental permission may be required depending on the nature of questions regarding mood. The IRB would want to consider how to handle subject reports of depressed or anxious mood.

The IRB would also consider whether completing a questionnaire asking questions about mood is detrimental to subjects who might be prone to depression? If so, parental permission would be required. The IRB might also require a school psychologist or counselor to be present to respond to any negative reactions by subjects. Subjects would then be told that a counselor is available to help subjects deal with any negative reactions to the study.

• Student researcher wants to investigate the relationship between SAT scores and GPA through peer's self report.

Minimal risk study: Parental permission not required if data are collected anonymously and subjects are informed that their participation is voluntary and that they can withdraw at any time.

• A student wants to show his classmates an optical illusion graphic and compare the responses of boys and girls.

Minimal risk study: The IRB would want to consider the nature of the optical illusion. Would anyone find it offensive? If not and the data are collected anonymously, parental permission could be waived.

The student researcher must provide information to the research subjects about what they will be asked to do, the voluntary nature of participation and their right to withdraw at any time.

- Do students do better memorizing words while listening to Mozart or rock music?

Potentially more than minimal risk study: The IRB would first want to know exactly what music was to be used. What if the rock music had profanity? Who determines the definition of profanity - the most conservative parent?

If the IRB determines that the music might be offensive (even slightly) to someone, parental permission should be required. The consent form should describe the music to be presented and give parents the opportunity to hear the music if he or she requests.

If the IRB determines that the music would not be offensive to anyone and the data are collected anonymously, they may waive the requirement of documentation of informed consent. However, the student researcher must provide information to the research subjects about what they will be asked to do, the voluntary nature of participation and their right to withdraw at any time.

- Do students who have math class in the morning do better on a test of "simple" math problems than those who have math class in the afternoon?

Potentially more than minimal risk study: The IRB must

determine the stress level associated with a "simple" math test. The committee might consult with both math teachers regarding the level of stress associated with the test for all students. If math teachers and IRB are comfortable with the "simple" math test not resulting in stress, the data are collected anonymously and the potential participants are not at risk for negative feelings related to the findings, the IRB could waive need for documentation of parental permission. However, some IRBs may require documentation of parental permission in this situation.

The student researcher must develop recruiting procedures that highlight that participation in the study is voluntary and that students can withdrawal from the study at any time. Efforts must also be taken to ensure that students that do not want to participate must be able to decline participation inconspicuously.

- A student wants to show elementary students an optical illusion graphic and compare the responses of boys and girls.

Minimal risk: As long as the optical illusion is not offensive to anyone, the study could be considered minimal risk and parental permission could be waived. However, some IRBs and school professionals may decide to require parental permission.

- Do children do better on a spelling test after listening to a certain type of music?

Minimal risk: The IRB should consider potential risks associated with whether some might find the music "offensive," or whether there is stress associated with taking a spelling test. Are there privacy and confidentiality issues? If the music was deemed to be innocuous, parental permission could be waived.

More than minimal risk: The IRB, school principal or teacher should require parental permission due to any reservations they have about the impact of the project on the subjects or parents' reaction to their child being part of a research project.

- Student researcher wants to know how fast boys and girls can run upstairs.

More than minimal risk: Documented parental informed consent required due to risk of injury. IRB might require safety precautions (e.g., a school nurse must be present, limit amount of stairs to 1 flight)

- Student researcher goes to the swim practice and times the swimmers as they are engaged in their regular swim practice (supervised by an adult coach)

Minimal risk: Student researcher is only observing. IRB may waive the need for parental permission because the swimmers are not being asked to do anything by the student researcher.

• Student researcher asks members of the swim team to participate in her study in which they have to swim 2 laps. This occurs after swim practice or on a day in which there is not practice

Potentially more than minimal risk: two possible options for IRB: 1) Require parental informed consent and require that a lifeguard present, 2) Instead of parental permission the swim coach gives the OK that swim team members are capable and the coach and/or lifeguard are present. In either case, the research subject must be informed directly that participation is completely voluntary and that he/she is free to stop participating in the project at any time.

• Student researcher wants to know if listening to rock music affects driving ability. He plans to test driving ability in the school parking lot with students driving their own cars around cones.

More than minimal risk: Requires documentation of parental permission for subjects and multiple safety precautions. The IRB may also require documentation that the school principal is aware of and approved the study. Many IRB's would not allow this project to be conducted because of school liability issues.

- Student researcher wants to know if listening to rock music affects driving ability. He plans to test driving ability with a video game.

No more than minimal risk: The IRB should listen to the proposed music and consider whether any parents would be take offense to the music. IRB would also want to consider the nature of the video game. IRB action may depend on the age of potential subjects (e.g., 6 graders vs. 12 graders). Different IRBs may come to different conclusions or different courses of action. IRBs that decide to waive parental permission in such situations may wish to document that the study was reviewed and approved by a principal or administrator.

Additional Resources

http://www.med.umich.edu/irbmed/FederalDocuments/hhs/HHS45CFR46.html
Code of federal regulations for the protection of human subjects
http://www.hhs.gov/ohrp/irb/irb_guidebook.htm
A guide produced by Office for Protection of Research Risk (OPRR) of the US Department of Health and Human Services (HHS). This resource can be used by IRBs to help them with their review. Includes an extensive appendix of additional resources.
http://www.nihtraining.com/ohsrsite/IRBCBT/intro.html
A computer based training course for new IRB members.

http://www.fda.gov/oc/ohrt/irbs/informedconsent.html
A guide to informed consent from the Food and Drug Administration"

©Society for Science and the Public – used by permission

Helpful Tips for a Great Science Fair Experience

You've come up with a fantastic idea, produced a dynamite project proposal, conducted months of design and experimentation, and achieved the highest grade in your school class. That, in and of itself, is a major achievement – but it's not enough to walk into your local Science Fair competition and win a top prize. In fact, some would say that, to paraphrase Winston Churchill, you've only reached "the end of the beginning."

The next critical steps will truly determine whether your groundbreaking project is well received by the judges, or even allowed to compete at all. Yes, it's true, each year there are many fantastic projects that may make it through the school level, and occasionally even through the regional fair, only to be disqualified from participation at the state level.

Here are some tips you may want to consider to achieve a truly amazing science fair competition. These all presume that you've already done a great job on your project itself, and that you can speak intelligently about each and every aspect of the experiments from beginning to end. We'll talk about preparing your verbal and visual presentations, how to conduct yourself during your interview, what to bring to the science far, and how to ensure that you actually have the opportunity to compete at all. Let's actually start with the last topic – disqualification. I had the opportunity to attend the 2011 science fair and saw a fantastic project (one that later went on to win first place in another national science fair) be disqualified from participating. Let's talk about some of the most common reasons why a project may fail to qualify; using information extrapolated, with permission, from the Society for Science and the Public, the parent organization of the Intel Science and Engineering Fair.

Preventing your project from disqualification

In the guidelines provided by SSP, they cite 7 specific risks of potential disqualification of projects, which are outlined as follows:

1. Studies involving Human, vertebrate animals, or potential hazardous biological agents, that lack the proper approval forms.
2. Prohibited vertebrate animal studies, which may include, but are not limited to experiments which cause pain or suffering, toxicity research, predator/prey, restriction of food and water, or those in which vertebrate animals were euthanized.
3. Studies involving microorganisms cultured at home, instead of in a suitable laboratory setting.
4. General eligibility issues such as more than three persons on a team project, a person competing individually when the work was that of a team, a

project more than one year in length.
5. Scientific misconduct, including plagiarism and falsification of data.
6. Continuation projects misrepresented as this years work.

Clearly each and every one of these issues can be easily avoided by paying meticulous attention to the rules and regulations, getting help where needed, and never being afraid to ask your advisor, or even your local science fair official to review your information well in advance of the competition itself.

Preparing your poster for presentation

At each science fair there are always one or two projects that stand out, if not on the basis of the research they have done, than on the sheer impact of their poster and display. Remember that you have space limitations, and it's vital to know those before you even begin the process, so that you don't find yourself with a 49-inch display at a science fair with a 48-inch width limit. Each fair has restrictions on the height, depth and width of the complete display, and they have no flexibility in these matters, so see your local science fair information sheet for their specific parameters.

Within the size limitations, students are free to exercise a tremendous amount of creativity in the design of the their personal display. For many students, this involves connecting two trifold poster boards together vertically, with bracing on the rear to hold them together. This creates a 72" high area for the application of whatever experimental information they wish to present. It does create the inherent problem of getting it around and making sure it doesn't fall apart in the middle of the science far – something that has happened many times.

An alternate method that has become popular over the last few years is the use of what is known as a pop-up display banner. These are professionally produced at a relatively low cost, based on a PDF file the student supplies to the graphic design house. Instead of using poster boards and attaching paper results to it, you prepare in Photoshop, pages or some other suitable graphics program, your full poster display, usually as a 65-inch high by 48 inch wide (if that is the allowable size at your fair) file, which you then convert to a PDF. With most of the graphic design companies, you just upload your PDF file right to them, and the print it in full color onto a canvas like material, which is then rolled up and placed in a spring loaded base. The functions much like the old spring tensioned pull down shades, except instead of pulling the shade down, you pull the banner up. With the aluminum base and rod structure, along with wide spread lower legs, the pop up banner is remarkably stable and very dramatic, offering the research incredible latitude in how they choose to present the project. These generally cost in the range of $150-$250, which is not inexpensive, but is seemingly insignificant when balanced against the many months you spent working on the project itself. For an example of these in use, you might check out some of the offerings at **www.popupstand.com**.

Of course, the poster itself is often just one part of the display – especially in a large biological or engineering project. Other aspects, which make up the display, are just as important, as they can result in your display being disallowed. For example, your display can have no living organisms (including plants), no taxidermy specimens, preserved vertebrate or invertebrate animals, food, bodily fluids, chemicals, and drugs. In addition, it is not permitted to display even photographs of vertebrate animals in laboratory procedures, or any stress situations. Furthermore, you would not be allowed to display anything made of glass, and chemical compounds, sharp objects such as needles, or engineering projects, which contain any exposed wiring.

When displaying photographs on your poster, each one must be credited to the photographer. If the experimenter took all of the photographs, then one photo credit at the bottom is sufficient. All photographs of human subjects must be accompanied by consent and release forms. The SSP website at www.societyforscience.org has a fantastic resource section showing examples of posters that meet the requirements, as well as those that did not. It's suggested that all potential researchers check out this section for some helpful tips.

Remember that the very first person you will see at the science fair is one of the most important, namely the "Safety Officer." He or she will need to completely sign off on all aspects of your display before you are allowed to even compete, so what for that person, be very respectful and prepared to explain not the science of the project, but rather the compliance of the display itself. Always bring a tape measure, just in case there are any questions as to whether your display is the correct size and configuration.

Dealing with the science fair judges
The men and women who volunteer their time to judge the science fairs represent as wide a range of potential people as you will ever meet. They are parents, corporate executives, and in many cases – especially at the higher levels – they are renowned science and engineering professionals in their own right. Without the help of these volunteers, the entire system would quickly grind to a halt, and as such, it's important to be respectful of the great time and energy they are bringing to the process itself. They will be viewing a wide array of projects the day you meet them at the fair, and each brings to the table their own unique perspective and experiences. Some will be far more knowledgeable about the subject of your experiment than you are. Others may have only a very limited understanding of what you have accomplished. One of the most important things to note, in either case, is that each judge has their own unique style, and one of the best things you can do is to ask them up front just how they would

prefer you to present. Some judges want to take a few minutes to look at your poster and get an understanding of just what you have done. Others focus less on the poster, and would prefer you to just tell them about the research. Either is completely fine, but it's vital to know whom you're dealing with. There are many good ways to try to establish this up front, but I have found that the easiest, by far, is to just come right out and ask them – "Would you like to take a moment to just review the project first?" If they say yes, just let them take their time, and when they are ready to interact, they will.

Another tip I have found helpful is to look at the name tag of the judge. Usually, they show both the name of the judge, and the company, if any, with which they are affiliated. As you are presenting your proposal on solar energy, and your judge walks up with a badge that says "Tim Moore – Evergreen Solar," you can assume that he is well versed in all aspects of the topic, and leave out the elementary discussion as to what a solar cell is made of. At the same time, your next judge may have a badge that says "Susan Johnson – DNA Sequencing Inc." You can assume that she is likely a brilliant genetic researcher, but may need you to explain the basics of a solar cell in more detail before jumping into your innovative new solar design.

The use of supplemental judging materials is allowed, and can be very helpful, both to the student scientist, and the judges. Each science fair issues a scoring rubric, which lays out the very specific number of points available in each category, and what is required to score the maximum on each of the items. The judge will be listening attentively to your presentation, asking questions, and interacting with you – but then they are going to leave, and fill out their judging sheet privately. At that point, will they completely remember where your project hit each of the bullet points on the rubric? To be certain, you can prepare a separate judging packet, which can be as simple as a copy of the rubric, followed by a brief synopsis of your project, with the heading of each page carrying the title of one of the target items in the rubric. In this way, when the judge leaves your project area, they can refer to the packet and see how you hit each and every one of the rubric items, rather than trying to recall if you had all of yours references cited, or all of your forms provided. The package can include a copy of your final project proposal, charts and diagrams with results and discussion, and anything else you would like them to have.

A great way to know what the judges are looking for, over and above the items expressed above, is to take a look at the "resources for judges" that are provided directly by the Society for Science & the Public, the parent company of the Intel Science and Engineering Fair. While this is, of course, just one of the many science fair competitions, and other fairs have different guidelines, it is one of the largest, and the guidelines are a great reflection of how proper judging should be done. As this is one of the most important aspects of your competition, it is presented here, almost in its entirety, again with the gracious permission of the SSP leadership. All information in the following section is ©SSP.

Guidelines for Judging
Interviews

The core of judging at the Intel ISEF is the interaction between judge and student in the interview. During these discussions, the quality of the project being presented is being evaluated. Just as importantly, the judge is serving as a STEM (science, technology, engineering, and math) ambassador. Judges must strive to make the student(s) comfortable so that they can demonstrate and share their knowledge and present their project to the best of their ability. In preparation for this important duty, judges are strongly encouraged to preview projects in their judging category in advance of personal interviews.

The unifying theme for Intel ISEF judging is: "Reward the best; encourage the rest." Judges are reminded to keep this notion in mind during their interaction with students.

Logistical and practical guidelines:
- Be prompt to start the interview and spend the entire 15-minute interview period in discussion with the student(s).
- Be friendly, courteous and professional.
- Demonstrate an interest in the student's work. Do not

criticize, dismiss or display boredom about any project, even if you consider it an unviable award candidate.
- Attempt to learn something from each project (no matter how unsophisticated it might appear).
- Distinguish among shyness, nervousness and lack of knowledge.
- If interviewing a team, draw out each member of the team for at least a portion of the interview. It is not appropriate for a single team member to serve as the exclusive spokesperson for the project.

You are encouraged to:
- Take control of the interview and do not allow the student to monopolize the interview with a canned speech.
- Get to the heart of what the student(s) has/have done and whether their work is worthy of an award relative to the other projects in the category ("Reward the best.")
- Remember that the display represents a summary and is secondary to the student's knowledge of the subject and his or her ability to explain their work.

Questions should focus on the specifics of the project:
Probe what the student knows about the subject in general.
Read the hypothesis/engineering problem, then go to the result to see if these match; discern how the Finalist went from hypothesis/problem to result.
Ask about recognized or perceived errors.

Ask why the student was interested in this subject.

Multi-year Projects/Continuations:
If the project is a multi-year effort, the Finalist is required to have an Intel ISEF Continuation Projects Form 7 visible at the project. However, please be aware that the ONLY work being judged and discussed in caucusing must have been completed in the current year research period.

What did the student actually do?
Judges often struggle with the distinction between the contributions of the student and the contributions (intellectual and resource) from other individuals (parents or mentors). This is especially true when the research is conducted at a Registered Research Institution.

For a project performed at a Registered Research Institution:
- Ask fundamental science or engineering questions to gauge the Finalist's understanding of the research conducted.
- Ask for specifics of what the student did, and what others did. How did his/her project fit into the ongoing work of the laboratory?
- In the preview of the project, review Form 1C: Registered Research Institutional or Industrial Setting Form. This form is completed by the mentor at the end of the research to articulate what the student did during the project.
- It is appropriate to hold projects to a higher standard

when they have benefited from this level of professional guidance. How did the student take advantage of the opportunity to work in a Registered Research Institution and positively use the resources available through the lab?

For a project that was performed at home or in a school laboratory:

- Can the student provide examples of the trials and errors that might be expected without professional guidance? How did the Finalist demonstrate his/her problem-solving ability?
- What help did the student receive? Being outside of a laboratory setting does not necessarily mean that professional guidance was not available or used.
- How did the student take advantage of the tools or resources available to them or overcome the absence of such resources?
- Student initiative and ingenuity may be taken into account in the scoring in the "creative ability" and "skill" categories. Students are judged on what they did, not what was done by others for them.

©Society for Science & the Public

Things to bring to the Science Fair

Giving an effective presentation has a lot to do with your mood and how you feel the day of the event. If you're starving, have a headache, and are worried about your AP Biology Test the following day, you're just not going to be able to give the effective presentation that will make your project shine. As you pack up your poster and supplies, you'll want to bring along a backpack with the following list of potential items:

A bag lunch, snacks, candy, granola bars, a few drinks, some Advil or Tylenol in case you get a headache, a box of tissues for when your nose starts to run during your presentation, extra duct tape for your project, an extension cord and power strip if your project needs electricity are all great things to have on hand. It also wouldn't hurt to bring an extra tie and shirt, so that when the drink you brought spills all over you, you have something to change into!

While the fair might easily be the most exciting thing to happen to you in a long time, it is interspersed with long periods of down time in between judging. You may go for two hours without seeing a single judge, and that can get really boring – especially since you are required to stay with your project at all times. Bring along your homework or a good book, your laptop, or even a deck of cards so you and your newfound friends can get to know each other a bit.

The Projects Themselves

As was discussed, projects go through a sequence of several different stages. Those who win the highest scores at their school, are then sent to compete in one of the regional fairs in their state. From that pool of highly competitive projects, the judges select the best of the best to compete at the state science fair.

The projects you will see on the following pages, all made it to the Massachusetts Science and Engineering Fair, held at M.I.T., in 2011. While I was competing at the fair myself, I went around and spoke to some of the other students, and asked if they would be willing to contribute their work to this book, in the hopes of guiding other students to success in future fairs. These experimenters were all incredibly gracious, and stepped up to do whatever they could to help the scientists of the future to hit the ground running.

Some editing was done to the projects you will see, mainly due simply to formatting requirements of the book. There were occasionally charts, photos and data tables that just were not loving being reduced from their original 8.5 x 11 format to a 6 x9 book. Paragraph indents, used by some, were replaced with spaces after paragraphs, and all fonts were normalized to 12pt Times, with 1.5 line spacing. This doesn't affect the main objective, which is to learn from these students not the specific results of a particular project, but rather the general thought process and formatting that enabled them to rise to the top of all other projects in 2011.

The Relationship Between Traffic and Stomata Density in Acer Platanoides

By Marouane Attioui

Abstract

The purpose of this experiment was to explore the relationship between stomata density and vehicle density. 25 leaves from each of the two locations were tested. The study was performed to better understand how Acer platanoides react to vehicle emissions, one of the bigger sources of pollution. A high traffic density location and a low traffic density location were the source of the leaves. The traffic densities for both places were counted around the same time of the day that the leafs were collected, but on different days. Many variables that affect stomata density were controlled. The high traffic density had an average of 1315 cars per hour while the low traffic density location had an average of 17.6 cars per hour. Stomata of Acer platanoides trees were counted under a compound light microscope. Based on the data, there was an inverse relationship between the vehicle density and the stomata density. Stomata densities for the high traffic location were averaged to be 45.65 stomata per 480 microns and the stomata densities for the low traffic density locations were averaged to be 68.84 stomata per 480 microns. Many reliable studies comparing CO_2 levels with stomata density done by Dr. Ian Woodward of Sheffield University and Dr. Robert B. Jackson of Duke University are consistent with my results. Since CO_2 outnumbers the other pollutants in vehicle emissions, it is most likely responsible for the decrease in stomata density. It seems that the trees have acclimated to the excess CO_2 and other gasses and pollutants by producing less stomata, thus taking in less CO_2 to keep homeostasis.

Introduction

Stomata complexes are mouth shaped pores located mainly on the underside of leafs and on roots. Each stoma complex consists of two guard cells that control the opening and closing of the pore (Torri, 2007). Each stoma complex also consists of subsidiary cells that cushion and supply the guard cells with water and ions (Sengbusch, 2003). Stomata are very important for the successful processes of photosynthesis and transpiration. For photosynthesis they are mainly responsible for the exchange of CO2 and H2O, CO2 enters and H2O leaves. For transpiration stomata allow some water molecules to escape, causing a negative pressure gradient that helps draw additional water up the roots (Wakefield, 2007). Stomata Density is the amount of stomata per area (in this experiment the area will be 480 microns).

Traffic Density is the amount of vehicles that pass through a road per hour. These vehicles release exhaust, also known as vehicle emissions. Traffic emissions consist of diverse gases and particles (EPA, "Automobile Emissions", pg. 2). They consist of gasses like nitrogen oxides, carbon monoxide, hydrocarbon gasses, and particulate matter. An annual test done by EPA showed that the amounts of carbon dioxide released by traffic greatly outnumbered all the other gasses (EPA,"Emission facts").

With these two concepts one may ask: what is the effect of traffic density on stomata density? The answer to this question can lead to a better understanding of how changes in the environment affect trees.

Hypothesis

The trees in the low traffic density location will have lower stomata density, because there would be less pollutants from vehicle emissions that clog and close stomata. The trees in the high traffic density location will have a higher stomata density as an adaptation to the pollutants from vehicle emissions that close and clog stomata.

Procedure

In this experiment 5 leaves from each tree of 10 trees were used. For a total of 50 leaves. 25 leaves were collected from a high traffic density (American Legion Highway) and 25 leaves were collected from a low traffic density (Jacobs Park). All the trees sampled were of the acer platanoides genus. All the trees were growing close to the sidewalk and they all grew on flat landscape. All the trees had a bark width that approximately ranged from 16 to 22 inches. All the leaves picked ranged from an approximate height of 2.5 inches to 4 inches and a width of 3 inches to 5 inches. All the leaves in this experiment were taken from the lower part of the tree, and were growing in the shade. All the leaves were picked during the same day and in the afternoon.

The leaves, nail polish, glass slides, tape and labels were used. Each leaf was polished on both sides of the back so that the polish patches were symmetrical. The polish patches were about
1cm x 1cm. The polish patches were fairly thin. The polish patches were left to dry for 30 minutes. The labels were made to keep all the slides in order. Each label was taped to the end of a two-inch piece of clear tape. Each piece of tape was taped to its corresponding polish patch and removed slowly so that the polish patch stuck to the tape. The pieces of tape were handled carefully so that there were no fingerprints on the center. Then the piece of tape containing the polish patch and the label was taped to a glass slide. This was done for all samples.

The slides of all the samples and a compound light microscope were used. Each slide was first viewed under 40x to find an area without any smudges. The lens was focused on the center of the nail polish patched patch for all the slides. Then the lens was moved to 100x. This was to choose an area without big veins. Then the lens was moved to 400x to count the number of stomata. This was done for all the slides. The same microscope was used throughout the experiment.

To get an approximate measure of the field of view, a clear

ruler and the microscope were used. The microscope was set to the 40x magnification lens. The ruler was placed in the diameter of the field of vision. The spaces between the lines were counted and the lines were also counted. The lines were divided by two and added to the number of spaces. This total number was multiplied by 1000 to equal 4800 microns for the 40x field of vision. The 4800 microns was then multiplied by .1 to yield 480 microns for the 400x field of vision. The 480 microns was used for the stomata density; the number of stomata per 480 microns.

To count the traffic density, two timers were used and two people were involved. One person stood at the sidewalk of American Legion Highway and the other stood at the sidewalk of Jacob's park. The counting from both people started at the same time and ended an hour later, using the timers to keep track of the time. The counting was done for different days. The traffic for each location was averaged. The average number was for the vehicle density: the number of vehicles per hour.

Results

To calculate the traffic density for the two locations, vehicles were counted for one hour, for five days. Then each of the two data sets was averaged to give 1315 vehicles/ hour for the high traffic density location and 17.6 vehicles/ hour for the low traffic density location. The results for this experiment were that for the high traffic density location, the stomata density was low and for the low traffic density, the stomata density was high. The two stomata densities for each leaf were averaged to get one number. This calculation gave 25 numbers for the low traffic density and 25 numbers for the high traffic density. Then each data set was averaged to get one number for the stomata density of high traffic density location and one number for the stomata density of low traffic density location. The average stomata density for the high traffic density location was 45.65 stomata/480 microns. The average stomata density for the low traffic density location was 68.84 stomata/480 microns. The difference between the two averages was 23.19. When this data was

$$t = \frac{\bar{x_1} - \bar{x_2}}{\sqrt{\frac{s_1^2}{n_1} + \frac{s_2^2}{n_2}}} \quad \begin{array}{l} \mapsto \\ \mapsto \\ \mapsto \end{array} \quad \frac{\text{difference between means}}{\frac{\text{variance}}{\text{sample size}}}$$

presented in percentages, 45.65 was added to 68.84 to equal 114.49. 45.65 (the average stomata density for high traffic density location) was divided by 114.49 to equal 40%. 68.84 (the average stomata density for low traffic density location) was divided by 114.49 to equal 60%. The percent difference between the two averaged stomata densities was calculated by 23.19 divided by 114.49 to equal 20%. The standard deviation for the high traffic density location was 9.0299 and for the low traffic density it was 10.2872.

A t-test was done using the formula:

If t>2.064 or t<-2.064 then there is a significant difference between the stomata density of the low traffic density and the stomata density location of the high traffic density location. The two S's are the two standard deviations of each of the two groups. The two N's were the sample sizes of each of the two groups. The two X's were the two means of the two groups. 45.65 (X2) was subtracted from 68.84 (X1) to equal 23.19 (the difference), 10.28725 (S1) was squared to equal 105.8275126 and divided by 25 (N1) to equal 4.233100504. 9.029904 (S2) was squared to equal 81.53916625 and then divided by 25 (N2) to equal 3.26156665. 4.233100504 was added to 3.26156665 to equal 7.494667154. 7.494667154 was square rooted to equal 2.737638974. The difference of the two sample means was divided by 2.737638974 to equal 8.470802842 which equaled the t value.

Discussion

The trees for the low traffic density location had a higher stomata density on average equal to 68.84 stomata/480 microns, then those of the high traffic density location which was equal to 45.65 stomata/480 microns. No reliable experiments, comparing traffic density and stomata density were found. There were other experiments that compared CO_2 levels to stomata density; they all had the same results, like the experiment done by Ian Woodward of Sheffield University and the experiment done by Robert B. Jackson of Duke University. As CO_2 levels increased the stomata density decreased, and as the CO_2 levels were decreased the stomata density increased. Knowing that CO_2 gas greatly outnumbers all the other gases and particles produced by vehicle emissions, the decrease of stomata in high vehicle density location is greatly affected by the CO_2. Since excess CO_2 heats plants beyond their normal temperatures. The less stomata in high traffic density locations seems to be an adaptation that the trees have made to take in less CO_2. Having less stomata will decrease the amount of water loss and keep the plants in homeostasis. The trees in the low traffic density location had more stomata, to take in more CO_2 since it is most likely that there would be less CO_2 in a low traffic density, and there would be more competition between the trees. It is most likely that other gasses that make up vehicle emissions are also affecting the stomata density. The t-test gave a number equal to the t-value of 8.470802842 which is much higher than 2.064, thus meaning that there is evidence that the two data sets are significantly different. During the viewing of the samples under the compound light microscope many smudges of pollution were seen on the trees of the high traffic density location and not many pollution smudges were seen on the low traffic location. There were no gene mutations found on the samples. The results from this experiment give very important data, this data would be very important to a scientist studying the affects of pollution on plants, or how plants and animals are reacting to the change that we are making in the environment. This would help support the idea that our planet is facing global warming. Traffic emissions contain other gasses besides carbon dioxide. They contain gasses like nitrogen oxides, carbon monoxide, particulate matter, and hydrocarbon gasses. All of these gasses affect the stomata in bad ways. For example nitrogen oxides cause stomata to stay closed by stopping the flow of potassium ions through potassium channels. Another example is that carbon monoxide is known to cause stomata to stay half way open. Particulate matter clogs the opening of the stoma. While I was counting stomata under the compound light microscope, I saw what looked like particulate matter, but as I changed the focus I realized that it was only shadows. To be able to count the number of closed stomata and get a valuable number, I will polish the leaves before pulling them off the trees. To see if there is particulate matter, I would need an electron microscope. To improve the experiment the experiment, trees from location with zero vehicle density would also be sampled. There would be more trees sampled from each location. I would also sample a non-invasive species.

Sources and References

"Acer Platanoides ; Norway Maple." Docstoc – Documents, Templates, Forms, Ebooks, Papers & Presentations. Web. 05 Feb. 2011. <http://www.docstoc.com/docs/43750242/Acer-platanoides--Norway-Maple>.

"Acer Platanoides, Norway Maple." Invasiveplantsmi.org. 28

June 2008. Web. 5 Feb. 2011.
<http://www.invasiveplantsmi.org/mpias/MIPC%20Acer%20platanoides%20summary.pdf>.

Brand, Dr. Mark H. "Acer Platanoides." Acer Platanoides. Acer Platanoides, Autumn 2008. Web. 17 Dec. 2010. <http://www.hort.uconn.edu/plants/a/acepla/acepla1.html>.

Bunce, James A. "Carbon Dioxide Effects on Stomatal Responses." Ars.usda.gov. 14 Oct. 2003. Web. 18 Dec. 2010. <http://www.ars.usda.gov/SP2UserFiles/ad_hoc/12755100FullTextPublicationspdf/Publications/bunce/carbondioxidestomatal.pdf>.

"Car Exhaust, Air Pollution"; Combustion Engines: Carbon Dioxide; Nitrogen Gases." Alpha Online, Clean Air, Nutrition for the 21st Century, Solutions for Medical Problems. Alpha Nutrition/ Environmed Research Inc., 2 Dec. 2006. Web. 11 Dec. 2010.
<http://www.nutramed.com/environment/carsepa.htm>.

Dahlgren, Tami. "Health Affects from Automobile Emissions." Ecy.wa.gov. TDD. Web. 17 Dec. 2010. <http://www.ecy.wa.gov/pubs/0002008.pdf>.

"Emission Facts: Average Annual Emissions and Fuel Consumption for Passenger Cars and Light Trucks | Consumer Information | US EPA." Epa.com. 14 Aug. 2007.

Web. 05 Feb. 2011. <http://www.epa.gov/oms/consumer/f00013.htm>.

EPA. "Automobile Emissions": an Overview." EPA.gov, 12 July 2006. Web. 11 Dec. 2010. <http://www.epa.gov/otaq/consumer/05-autos.pdf>.

Fletcher, Ben. "Stomata Control How the Atmosphere Affects Plants." Shef.ac.uk. The University of Sheffield. Web. 18 Dec. 2010. <http://www.shef.ac.uk/aps/apsrtp/fletcher-ben/atmosphere.html>.

"Guard Cells." Penn State Biology Department, 23 Mar. 2001. Web. 18 Dec. 2010. <http://homes.bio.psu.edu/people/faculty/Assmann/lab/guardcells.html>.

Hewitson, John, Richard Price, and Roger Delpech. "What Are the Most Recent Theories on Stomatal Opening and Closure?" Science and Plants for Schools. SAPS, 24 Mar. 2010. Web. 19 Dec. 2010. <http://www-saps.plantsci.cam.ac.uk/records/rec106.htm>.

Kantharaj, G. R. "Plant Hormones-ABSCISINS." Plant Cell Biology For Masters- by G. R. Kantharaj. Web. 19 Dec. 2010. <http://plantcellbiology.masters.grkraj.org/html/Plant_Growth_And_Development6-Plant_Hormones-Abscissins.htm>

Kennesaw University. "Vehicle Emissions." Kennesaw University, 16 Feb. 2008. Web. 10 Dec. 2010. <http://esa21.kennesaw.edu/activities/smog-cars/doe-veh-pollutants.pdf>.

Kimball, J. "Gas Exchange in Plants." Users.rcn.com/jkimball.ma.ultranet. Kimball, 4 Nov. 2010. Web. 18 Dec. 2010. <http://users.rcn.com/jkimball.ma.ultranet/BiologyPages/G/GasExchange.html>.

Mauseth, J. "Protoderm, Promeristem." Web.biosci.utexas.edu. University of Texas. Web. 19 Dec. 2010. <http://www.sbs.utexas.edu/mauseth/weblab/webchap6apmer/6.2-2.htm>.

Mersie, Wondi. "Trees for Problem Landscape Sites -- Air Pollution - Virginia Cooperative Extension." Publications and Educational Resources - Virginia Cooperative Extension. 1 May 2009. Web. 17 Dec. 2010. <http://pubs.ext.vt.edu/430/430-022/430-022.html>.

Morison, James L. "Guard Cell Photosynthesis." Plant Physiology Online. University of Essex, Sept. 2006. Web. 18 Dec. 2010.
<http://4e.plantphys.net/article.php?ch=&id=265>.

Pillitari, Lynn. "Pillitteri Lab." Www.wwu.edu. Western Washington University. Web. 19 Dec. 2010. <http://faculty.wwu.edu/pillitl/Welcome.html>.

Roach, John, and David Dr. Peterson. "Ozone Damages Plants." Health And Energy Company. University of Washington, 8 Dec. 1999. Web. 18 Dec. 2010 <http://healthandenergy.com/ozone_damages_plants.htm>. Pollock, M. SpringerLink.com. 10 Sept. 2004. Web. 18 Dec. 2010. <Inhibition of stomatal opening in sunflower leaves by carbon monoxide, and reversal of inhibition by light>.

Schlageter, Martin. "Air Pollution - Pollutants & Health Effects." Coalition for Clean Air. Network Solutions LLC, 19 Mar. 2010. Web. 17 Dec. 2010. <http://www.coalitionforcleanair.org/air-pollution-pollutants.html>.

Sengbusch, Peter V. "Dermal, Parenchyma and Assimilation Tissues." Biologie.uni-hamburg.de. Impressum, 31 July 2003. Web. 18 Dec. 2010. <http://www.biologie.uni-hamburg.de/b-online/e05/05.htm>.

Sibley, David. The Sibley Guide to Trees. New York: Alfred A. Knopf, 2009. pg. 337. Print.

"Smog and Particulates: Pollutant Descriptions." Scorecard

Home. GetActive Software, 18 Oct. 2006. Web. 11 Dec. 2010. <http://www.scorecard.org/env-releases/cap/pollutant-desc.tcl>.

Torii, Keiko U. "Stomatal Patterning and Guard Cell Differentiation." Washington.edu. Western Washington University, 2007. Web. 19 Dec. 2010. <http://faculty.washington.edu/ktorii/Torii_PCM.pdf>.
Torii, Keiko U. "Stomata." UW Faculty Web Server. Western Washington University, 2008. Web. 19 Dec. 2010. <http://faculty.washington.edu/ktorii/stomata.html>.

"Transpiration - Major Plant Highlights." Library of Crop Technology Lessons. Web. 18 Dec. 2010. <http://croptechnology.unl.edu/viewLesson.cgi?min=1&max=8&topic_order=5&LessonID=109285384>.

Wakefield, R. "Leaves and Stomata." Pima.edu. Pima, 31 Oct. 2007. Web. 19 Dec. 2010. <http://dtc.pima.edu/blc/1003rded/004/004_Leaves_Stomata/stomata.html>.

Whiting, David. "Plant Physiology: Transpiration, Photosynthesis, Respiration." Cmg.colostate.edu. Colorado State University, Aug. 2010. Web. 19 Dec. 2010. <http://www.cmg.colostate.edu/gardennotes/141.pdf>.

Woodward, Ian. "The Influence of CO_2 Concentration on

Stomatal Density - WOODWARD - 2006 - New Phytologist." Wiley Online Library. University of Sheffield, 21 July 1995. Web. 05 Feb. 2011. <http://onlinelibrary.wiley.com/doi/10.1111/j.1469-8137.1995.tb03067.x/pdf>.

Green grass grows all around
by Allison Baker and Katelyn Cappello

Abstract

This project is the result of an experiment on the effect of pH solutions on plant growth, specifically grass. We changed the pH of the solution we used to water each plant. The purpose of this project was to determine if acidic, neutral, or basic solutions would affect the plant's growth. During this experiment we saw significant differences between the growths of the different pH groups.

In our project we first wanted to use the extremes of the pH scale, 1 and 14, but found out that those would be extremely harmful. We then decided to use the pH levels of: 2, 3, 5, 7, 9, 10, and 12. We researched online to find how to create each pH level. For a pH of 2 we used lemon juice due to the fact that we could not use water and vinegar to get a pH level lower than 3. For a pH of 3 we used straight vinegar. For a pH of 5 we used water with a small amount of vinegar. For a pH of 7 we used water and some baking soda because the tap water was not pure, it was a bit acidic, and we needed it to be a pure pH of 7 for our control group. For a pH of 9 we used mostly baking soda with only a little bit of water. For a pH of 10 we used water with bleach and for a pH of 12 we used straight bleach because there is no other way to obtain a pH of 12. To make sure these pH values were accurate we tested them with pH strips. Once we made the pH solutions we watered the plants every other day, unless the soil was moist, for six weeks.

Through our experiment we concluded that three of the pH levels showed some sign of growth. These pH levels were 5, 7, and 10. The pH of 5 had an average of 103 blades, while pH of 7 had 78 blades and ph of 10 had only 1 blade. Also, the plant that received a pH of 5 blades seemed much stronger than the pH of 7 and did not seem to be wilting as much. In the pH of 10 a few blades grew but then died.

From these results we formed the following conclusions. When comparing these three levels the one that showed the most growth and strongest blades was a pH of 5. The pH of 7, which is pure water, showed significant growth but began to become weak, started to wilt, and its growth slowed. The pH of 10 showed minimal growth. In one of our cups it had two blades, but they then died after three weeks. This project and its results could benefit farmers, gardeners, and landscapers.

Introduction

While gardening, there are several things that have an effect on a plant's growth. The things that effect the plant's growth are called growth influencing factors. The growth influencing factors include: time, temperature, humidity, water, light, nutrition, pH, oxygen, carbon dioxide, and pests. It is important for gardeners to be informed about all the growth influencing factors that could possibly affect their plants as it only takes one factor to be out of range to cause a problem. Having the ability to control the growth influencing factors and keeping them in the ideal range for the specific plant will result in maximum growth. PH is the amount of hydrogen ions in a solution. The pH scale ranges from 0 to 14. On the pH scale, a pH of 7 is neutral, 0 to 7 is acidic, and 7 to 14 is basic. While growing plants, the preferred pH level of soil is 6.5. (Miller, 2001) Pure water is generally neutral with a pH of 7. When chemicals are mixed with water the pH can be changed to become more acidic or basic. Some common liquids that have a different pH value than water include vinegar, wine, lemon juice, beer, tomato juice, milk and seawater. Vinegar has a pH of 3, wine has a pH of 2.8 to 3.8, lemon juice has a pH of 2.5, beer has a pH of 4 to 5, tomato juice has a pH of 4, milk has a pH of 6.3 to 6.6, and seawater has a pH of 8.3.

The growth influencing factors which include time, temperature, humidity, water, light, nutrition, pH, oxygen, carbon dioxide, and pests effect a plant's growth. Time is a factor which you have no control over. Whenever growth influencing factors get out of range, plant growth stops and damage to the plant may occur. For example, if the temperature of the environment becomes too hot, the plant will wilt, have tissue damage, and eventually die.

Temperature is a factor which plays a very important role for indoor plants. The ideal temperature for plant growth is 70-75 degrees Fahrenheit. Plant growth will slow down in hot, humid conditions, and therefore it is crucial to keep the temperature in the ideal range. The ideal humidity for healthy plant growth is 50%, although plus or minus 10% would still allow for the plant to grow. Water is extremely important to plants. Water transports the nutrients throughout the plant. It is important to water plants with water that does not contain too much sulfur (water that smells like eggs) or iron (water that leaves rust colored stains). Always make sure your plant is receiving the perfect amount of water. Too much water could drown your plant; however, too little water could kill your plant or make it wilt. (Pimental, Berger, Filiberto, & Newton, 2004) Light is another factor that effects plant growth. Increasing the light in your garden also means increasing the temperature. It is estimated that seedlings need about 25 watts per square foot of lighting to develop properly. (Coomes & Allen, 2007)

The major nutrients that plants need to grow include nitrogen, phosphorus, and potassium. Nitrogen is used for green, leafy growth, phosphorus is needed to grow healthy roots, large flowers, and develop fruits, and potassium is needed for thick, strong, branches and stems. (Leger, Howe, Gurevitch, Woo, Hickman, Ashton, & Lerdau, 2007) Secondary nutrients include calcium, magnesium, and sulfur. Although these nutrients are not used in the quantities that the primary nutrients are, they are still important and used throughout the growing of the flowering process. Plants also need small amounts of micronutrients including boron, copper, iron, manganese, and molybdenum. PH is important in plant growth as it affects the availability of plant foods and prevents the spread of soil borne diseases. (Miller, 2001) The ideal pH level of soil for plants is 6.5. (Miller, 2001) Without oxygen, nutrients cannot be absorbed into the plant. Plant growth requires a large amount of carbon dioxide. As the amount of carbon dioxide is increased, so is the growth rate. Lastly, pests effect plant growth. Pests eat away at plants and cause them to die. The only real way of getting rid of pests is to use pesticides which should not be used on a consumable crop. Growth influencing factors play a major role on the plant's growth rate and are important to keep in ideal ranges.

There have been studies done in the past that have changed certain growth influencing factors. In one experiment, the pH of the water used to water the plant was changed. The results of the experiment concluded that the very acidic water ate away at the plants and the very basic water poisoned the plant. (Plants and pH, 2004) Most of these studies had varied results and were dependent on the type of plant.

The growth influencing factors play a major role on how healthy a plant will grow. The major growth influencing factors that can affect a plant's growth if they are varied include: time, temperature, humidity, water, light, nutrition, pH, oxygen, carbon dioxide, and pests. When changing these factors it could have a potentially positive or negative effect on the plant's growth. Gardeners, farmers, and landscapers will benefit from these experiments which change certain growth influencing factors. Being aware of the growth influencing factors and the effects they have on a plant's growth will help people grow healthy plants. Having a background on the factors will allow people to notice anything occurring that is not beneficial to the plant and find a solution to the problem. This experiment will provide useful information that will benefit plant growth.

Experimental Design

Green Grass Grows All Around: pH and its effects on plant growth

I. **Purpose**:
The purpose of this project is to test the effect of different pH solutions on plant growth.

II. **Hypothesis**: If the grass seed plants are given acidic liquid rather than pure water, then the plants will grow taller.

III. **Materials-**
- Grass Seed
- Soil
- Plastic cups
- Water

- Lemon Juice
- Vinegar
- Baking Soda
- Bleach
- pH strips
- Measuring cups
- Measuring spoons
- Metric Ruler

IV. **Procedure**:
1. Place 1/2 cup of potting soil into each plastic cup.
2. Sprinkle ¼ teaspoon of grass seed to top of each plastic cup and lightly cover with soil and place all plastic cups in the same location receiving equal light.
3. Gently water with pure water (pH of 7) until soil is moist. (2 tablespoons)
4. Make solutions of varying pH values by adding vinegar or lemon juice to make acidic solutions and baking soda or bleach to make basic solutions. Test solutions with the pH strips to ensure accurate pH measurements.
5. Label the plastic cups as to which solution they will receive.
6. Continue to lightly water each plastic cup once a day with the specified solution. (1 tablespoon) Change what day to water, as it becomes visible that the soil is moist or dry.

7. Record observations weekly. (Record height, color, and appearance)

IV. Data/Results pH 2 charts-

# of blades	Plant 1	Plant 2	Plant 3
Week 1	0	0	0
Week 2	0	0	0
Week 3	0	0	0
Week 4	0	0	0
Week 5	0	0	0
Week 6	0	0	0

Shortest blade	Plant 1	Plant 2	Plant 3
Week 1	0 cm	0 cm	0 cm
Week 2	0 cm	0 cm	0 cm
Week 3	0 cm	0 cm	0 cm
Week 4	0 cm	0 cm	0 cm
Week 5	0 cm	0 cm	0 cm
Week 6	0 cm	0 cm	0 cm

Tallest blade	Plant 1	Plant 2	Plant 3
Week 1	0 cm	0 cm	0 cm
Week 2	0 cm	0 cm	0 cm
Week 3	0 cm	0 cm	0 cm
Week 4	0 cm	0 cm	0 cm

Week 5	0 cm	0 cm	0 cm
Week 6	0 cm	0 cm	0 cm

Plants with ph of 3

Number of blades	Plant 4	Plant 5	Plant 6
Week 1	0	0	0
Week 2	0	0	0
Week 3	0	0	0
Week 4	0	0	0
Week 5	0	0	0
Week 6	0	0	0
Shortest blade	Plant 4	Plant 5	Plant 6
Week 1	0 cm	0 cm	0 cm
Week 2	0 cm	0 cm	0 cm
Week 3	0 cm	0 cm	0 cm
Week 4	0 cm	0 cm	0 cm
Week 5	0 cm	0 cm	0 cm
Week 6	0 cm	0 cm	0 cm
Tallest blade	Plant 4	Plant 5	Plant 6
Week 1	0 cm	0 cm	0 cm
Week 2	0 cm	0 cm	0 cm
Week 3	0 cm	0 cm	0 cm
Week 4	0 cm	0 cm	0 cm
Week 5	0 cm	0 cm	0 cm
Week 6	0 cm	0 cm	0 cm

Plants with ph of 5

Number of blades	Plant 7	Plant 8	Plant 9
Week 1	10	12	11
Week 2	53	58	56
Week 3	65	74	74
Week 4	90	106	83
Week 5	92	108	85
Week 6	112	110	88
Shortest blade	Plant 7	Plant 8	Plant 9
Week 1	0.3 cm	0.2 cm	0.1 cm
Week 2	0.4 cm	0.1 cm	0.3 cm
Week 3	0.6 cm	0.5 cm	0.5 cm
Week 4	0.8 cm	0.4 cm	0.7 cm
Week 5	0.6 cm	1 cm	0.5 cm
Week 6	0.7 cm	0.3 cm	0.6 cm
Tallest blade	Plant 7	Plant 8	Plant 9
Week 1	0.8 cm	1.4 cm	1.3 cm
Week 2	5.8 cm	6.1 cm	7.2 cm
Week 3	8.2 cm	8.2 cm	8.6 cm
Week 4	8.9 cm	11.1 cm	8.5 cm
Week 5	11.4 cm	13.2 cm	10.4 cm
Week 6	13.5 cm	13.4 cm	10.5 cm

Plants with ph of 7

Number of blades	Plant 10	Plant 11	Plant 12
Week 1	4	14	10
Week 2	28	56	48
Week 3	38	70	72
Week 4	30	70	80
Week 5	35	83	103
Week 6	43	85	105

Shortest blade	Plant 10	Plant 11	Plant 12
Week 1	0.2 cm	0.1 cm	0.1 cm
Week 2	0.5 cm	0.3 cm	0.4 cm
Week 3	0.5 cm	0.1 cm	1 cm
Week 4	0.6 cm	0.2 cm	0.3 cm
Week 5	0.5 cm	0.6 cm	0.5 cm
Week 6	0.6 cm	1 cm	1 cm

Tallest blade	Plant 10	Plant 11	Plant 12
Week 1	1.2 cm	1.3 cm	2 cm
Week 2	6.3 cm	7 cm	8 cm

Plants with ph of 10

Number of blades	Plant 16	Plant 17	Plant 18
Week 1	0	0	0
Week 2	0	1	1

Week 3	0	1	1
Week 4	0	2	1
Week 5	2	0	1
Week 6	2	0	1
Shortest blade	Plant 16	Plant 17	Plant 18
Week 1	0 cm	0 cm	0 cm
Week 2	0 cm	0.1 cm	0.8 cm
Week 3	0 cm	0.1 cm	0.4 cm
Week 4	0 cm	0.1 cm	0.6 cm
Week 5	0.9 cm	0 cm	0.7 cm
Week 6	1.2 cm	0 cm	0.7 cm
Tallest blade	Plant 16	Plant 17	Plant 18
Week 1	0 cm	0 cm	0 cm
Week 2	0 cm	0.1 cm	0.8 cm
Week 3	0 cm	0.1 cm	0.4 cm
Week 4	0 cm	0.1 cm	0.6 cm
Week 5	1.2 cm	0 cm	0.7 cm
Week 6	1.7 cm	0 cm	0.7 cm

(editors note – some tables truncated)

Plants with ph of 12

number of blades	Plant 19	Plant 20	Plant 21
Week 1	0	0	0
Week 2	0	0	0
Week 3	0	0	0

Week 4	0	0	0
Week 5	0	0	0
Week 6	0	0	0
Shortest blade	Plant 19	Plant 20	Plant 21
Week 1	0 cm	0 cm	0 cm
Week 2	0 cm	0 cm	0 cm
Week 3	0 cm	0 cm	0 cm
Week 4	0 cm	0 cm	0 cm
Week 5	0 cm	0 cm	0 cm
Week 6	0 cm	0 cm	0 cm
Tallest blade	Plant 19	Plant 20	Plant 21
Week 1	0 cm	0 cm	0 cm
Week 2	0 cm	0 cm	0 cm
Week 3	0 cm	0 cm	0 cm
Week 4	0 cm	0 cm	0 cm
Week 5	0 cm	0 cm	0 cm
Week 6	0 cm	0 cm	0 cm

V. **Analysis**:

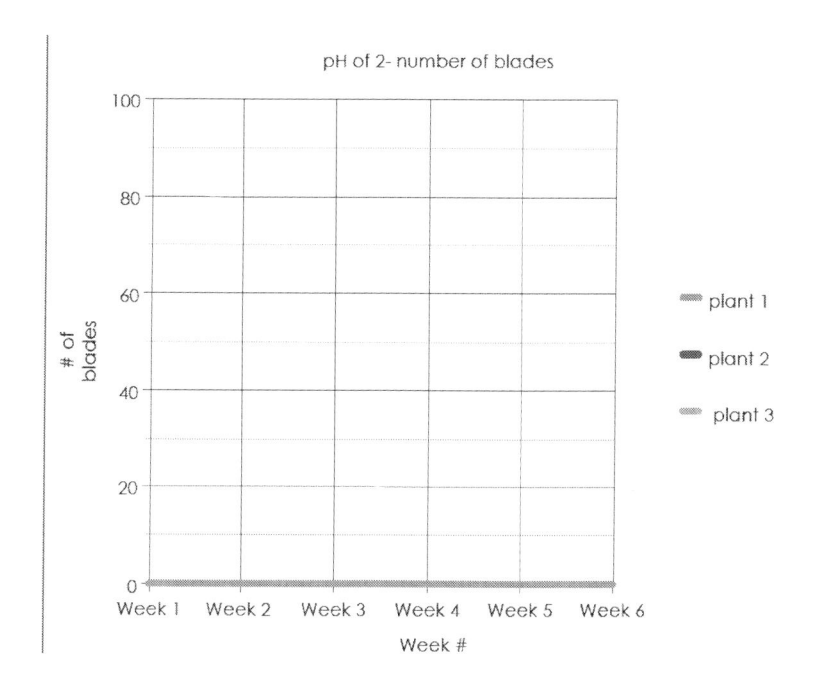

pH of 2 - number of blades

- plant 1
- plant 2
- plant 3

pH of 2 - shortest blade

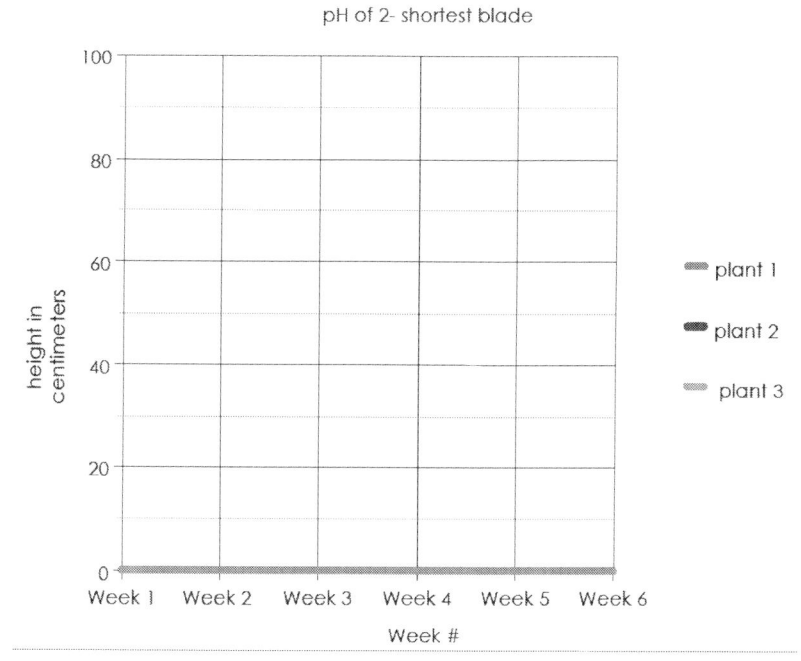

- plant 1
- plant 2
- plant 3

pH of 2- tallest blade

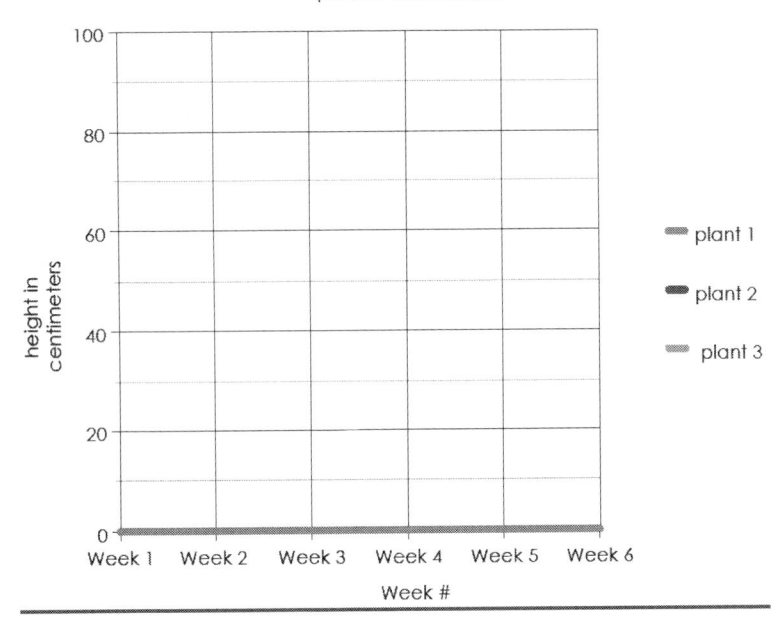

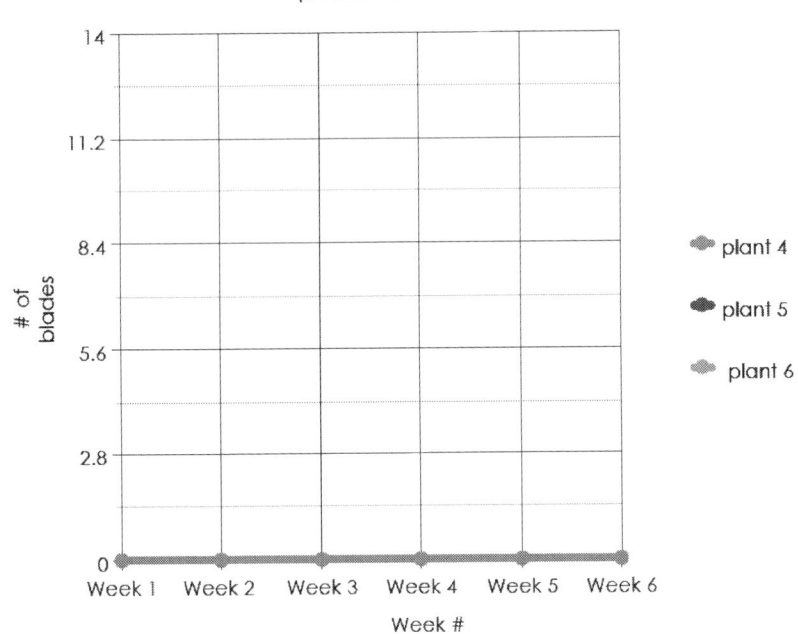

- plant 1
- plant 2
- plant 3

pH of 3- number of blades

- plant 4
- plant 5
- plant 6

pH of 3- shortest blade

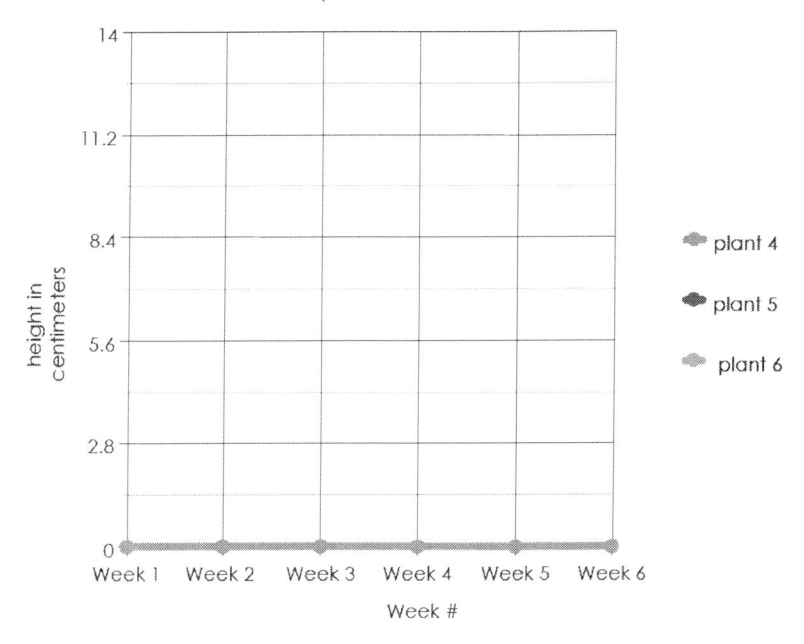

- plant 4
- plant 5
- plant 6

pH of 3- tallest blade

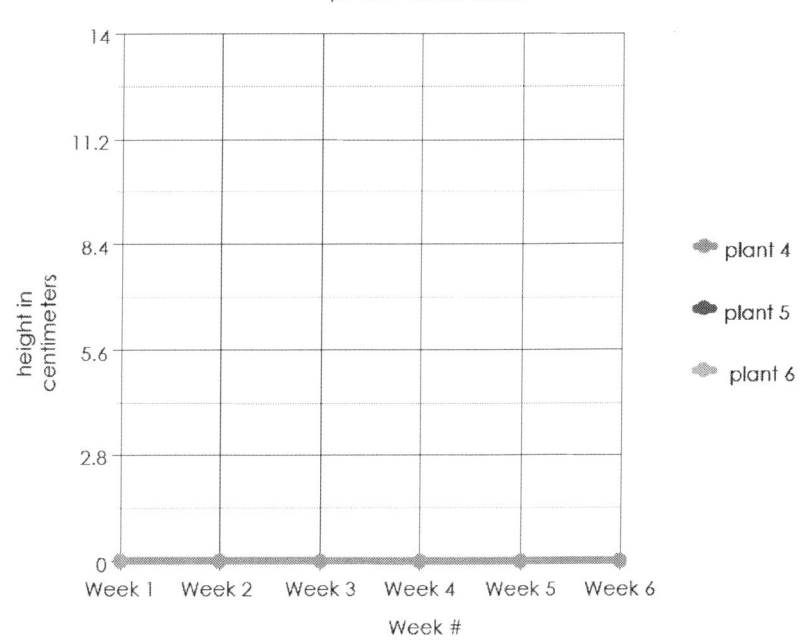

- plant 4
- plant 5
- plant 6

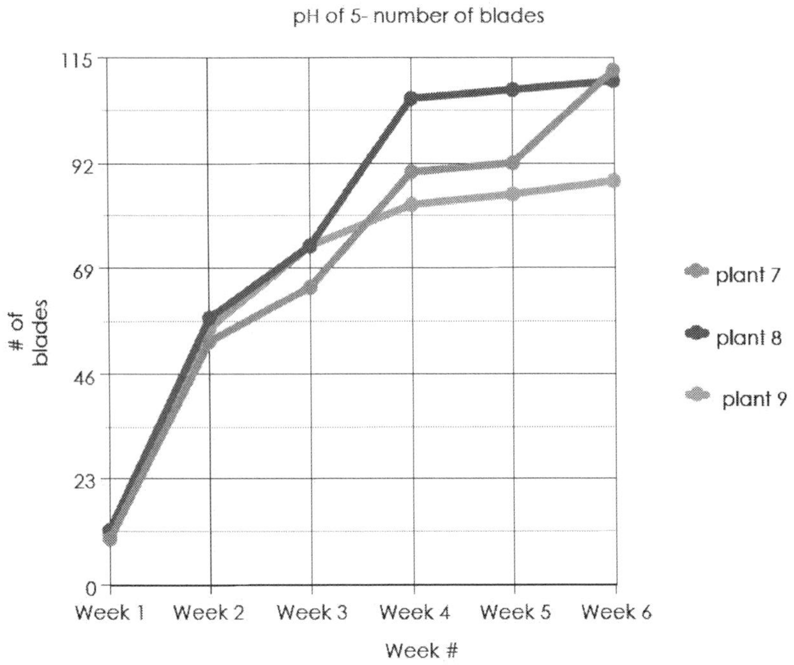
pH of 5- number of blades
115
92
69
46
23
0
# of blades
Week 1
Week 2
Week 3
Week 4
Week 5
Week 6
Week #
plant 7
plant 8
plant 9

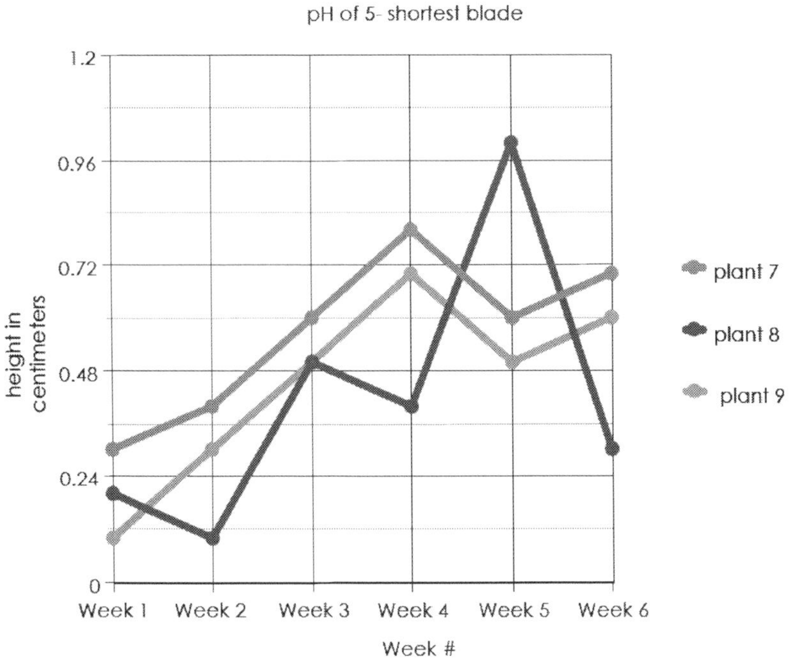
pH of 5- shortest blade
1.2
0.96
0.72
0.48
0.24
0
height in centimeters
Week 1
Week 2
Week 3
Week 4
Week 5
Week 6
Week #
plant 7
plant 8
plant 9

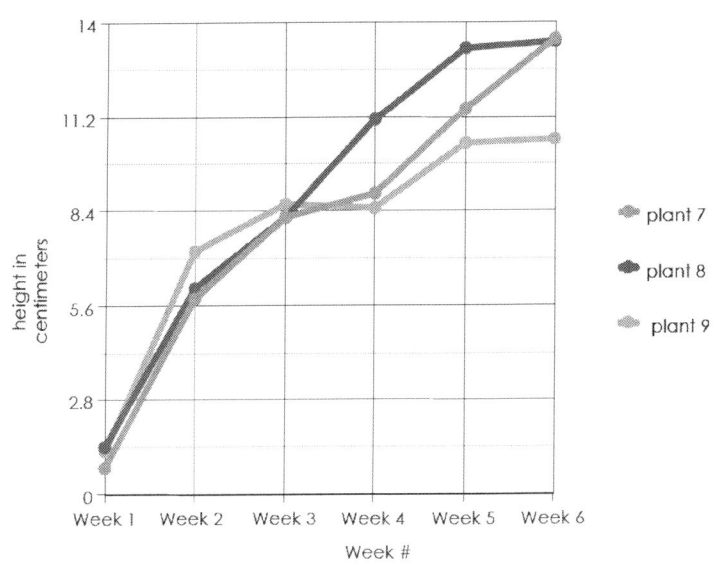

pH of 5- tallest blade

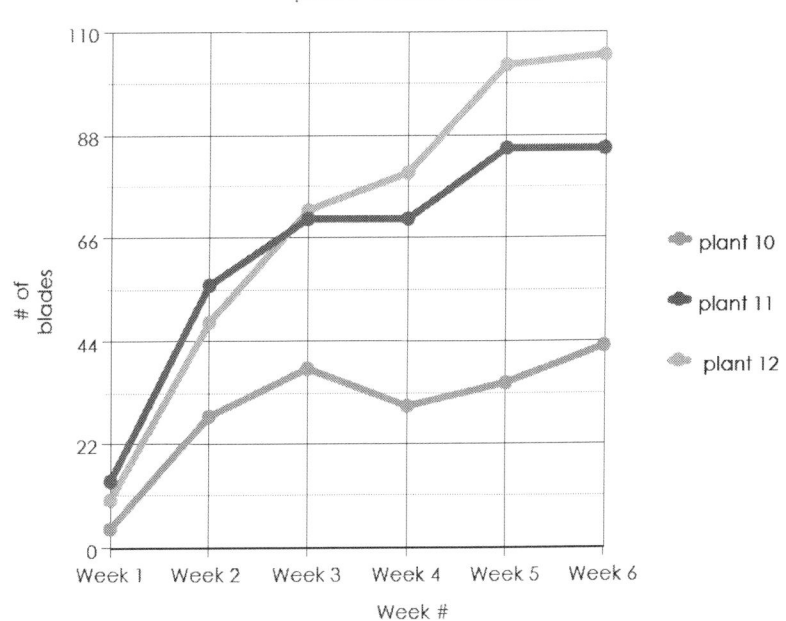

Chart showing height in centimeters vs Week # for plant 7, plant 8, and plant 9 across Weeks 1-6.

pH of 7- number of blades

Chart showing # of blades vs Week # for plant 10, plant 11, and plant 12 across Weeks 1-6.

pH of 7- shortest blade

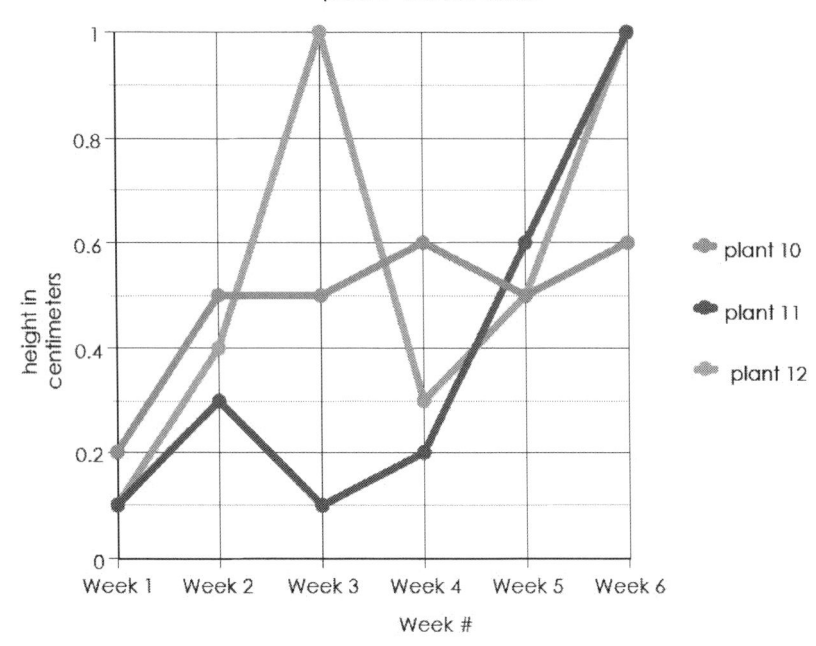

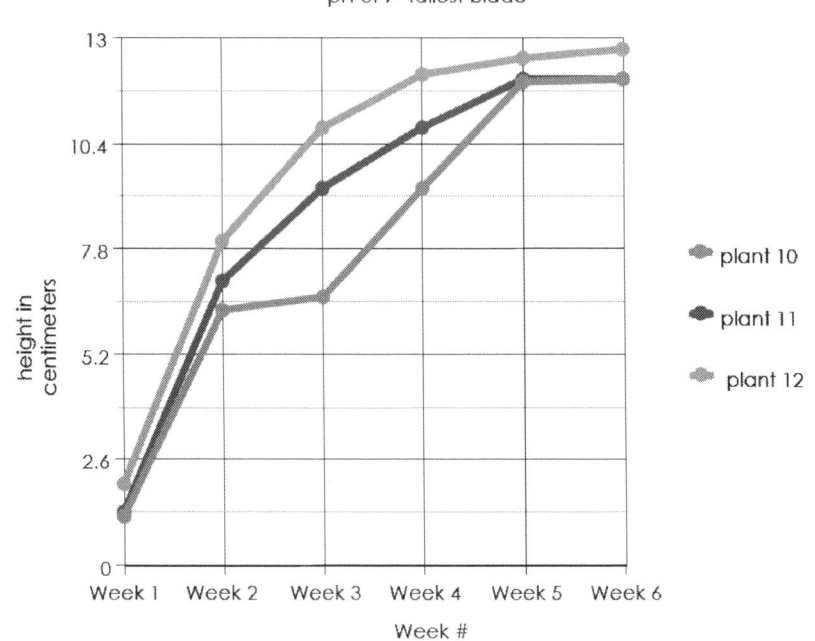

pH of 7- tallest blade

pH of 9 - number of blades

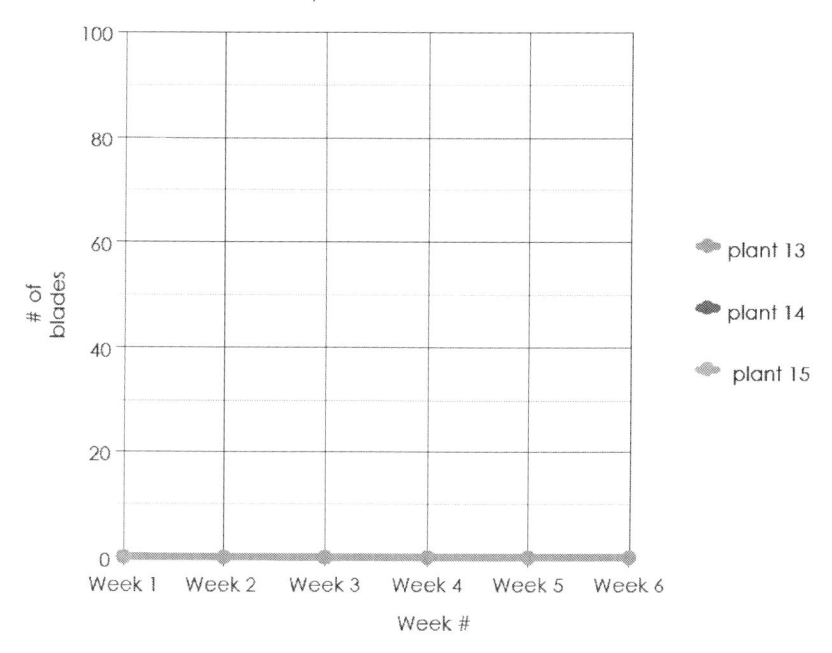

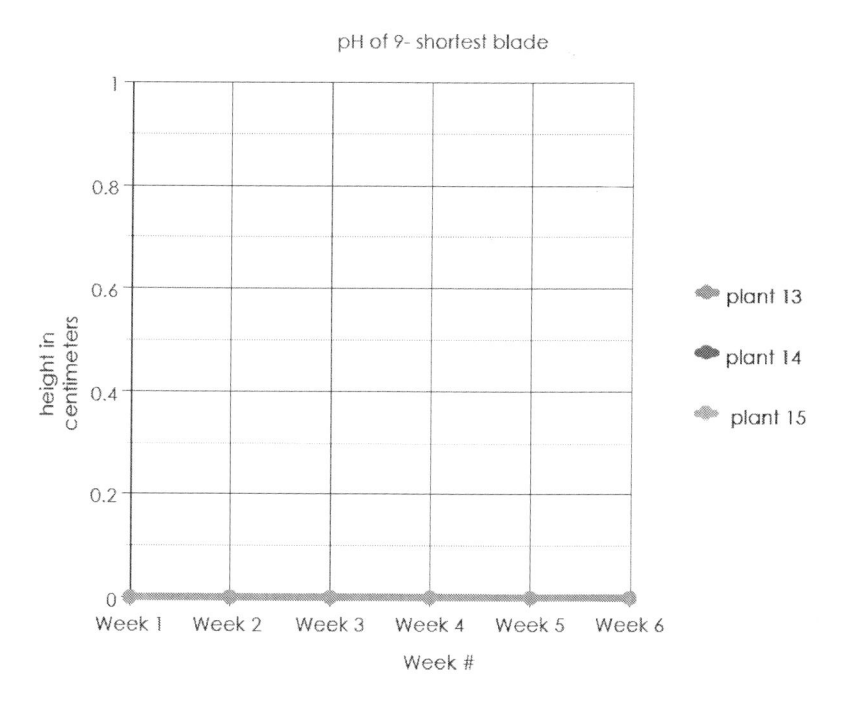

pH of 9 - shortest blade

pH of 9- tallest blade

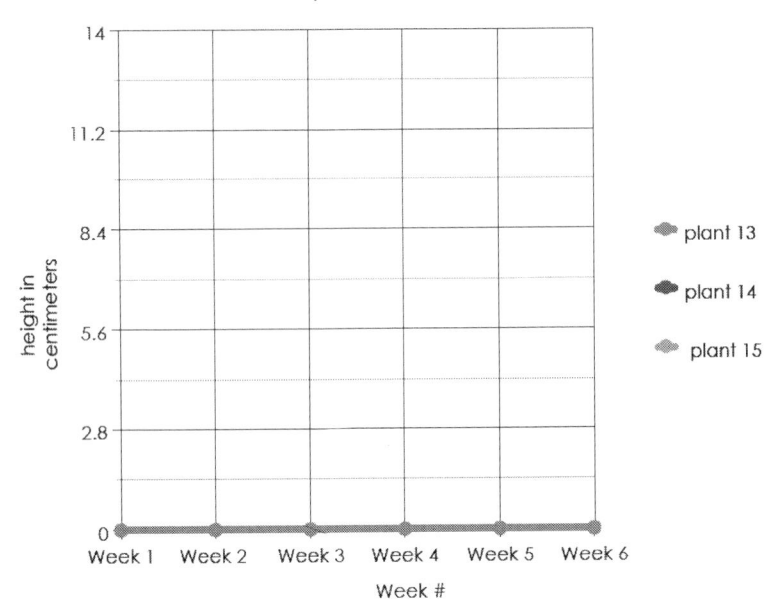

pH of 10- number of blades

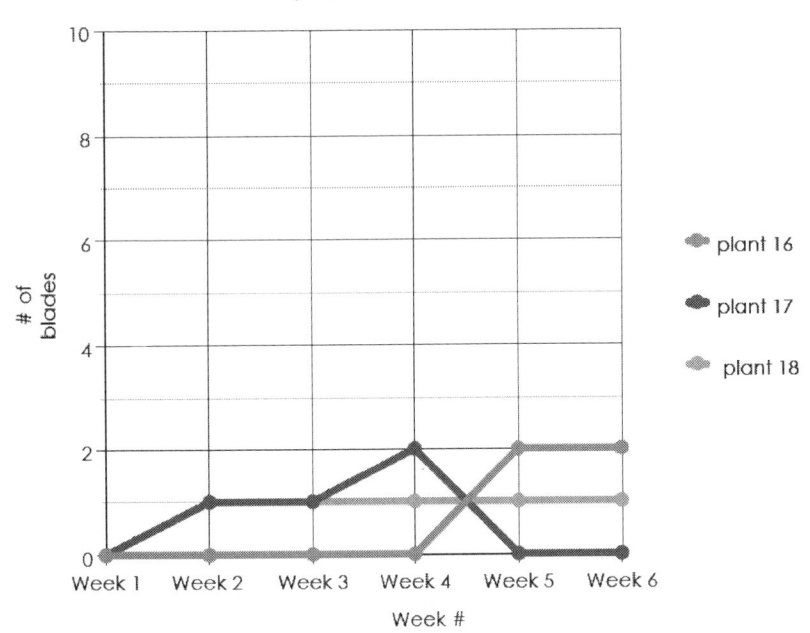

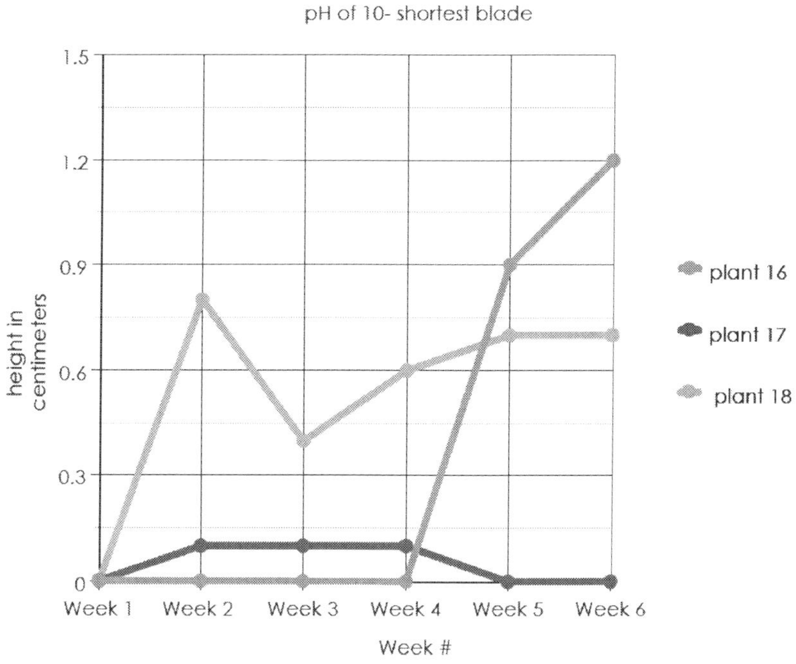
pH of 10- shortest blade
1.5
1.2
0.9
0.6
0.3
0
height in centimeters
Week 1
Week 2
Week 3
Week 4
Week 5
Week 6
Week #
plant 16
plant 17
plant 18

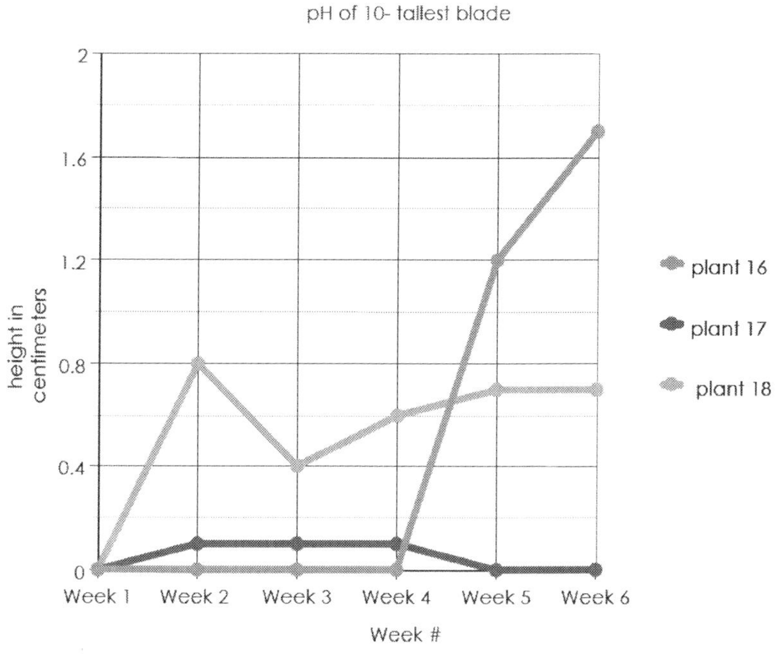
pH of 10- tallest blade
2
1.6
1.2
0.8
0.4
0
height in centimeters
Week 1
Week 2
Week 3
Week 4
Week 5
Week 6
Week #
plant 16
plant 17
plant 18

pH of 12- number of blades

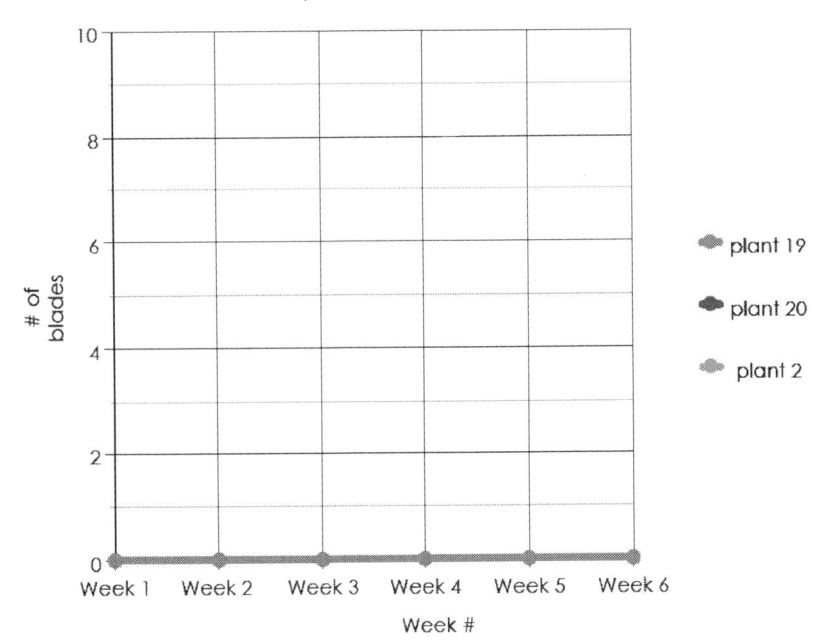

- plant 19
- plant 20
- plant 2

pH of 12- shortest blade

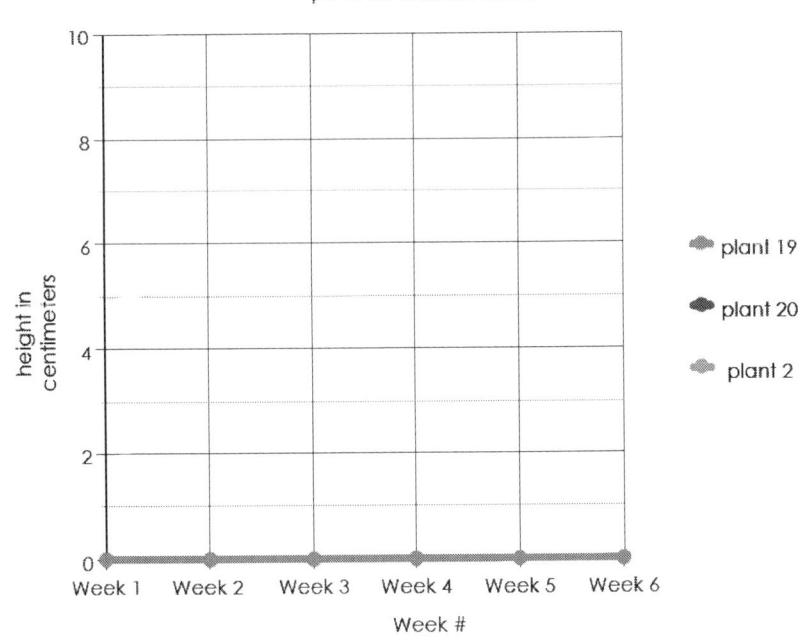

- plant 19
- plant 20
- plant 2

pH of 12- tallest blade

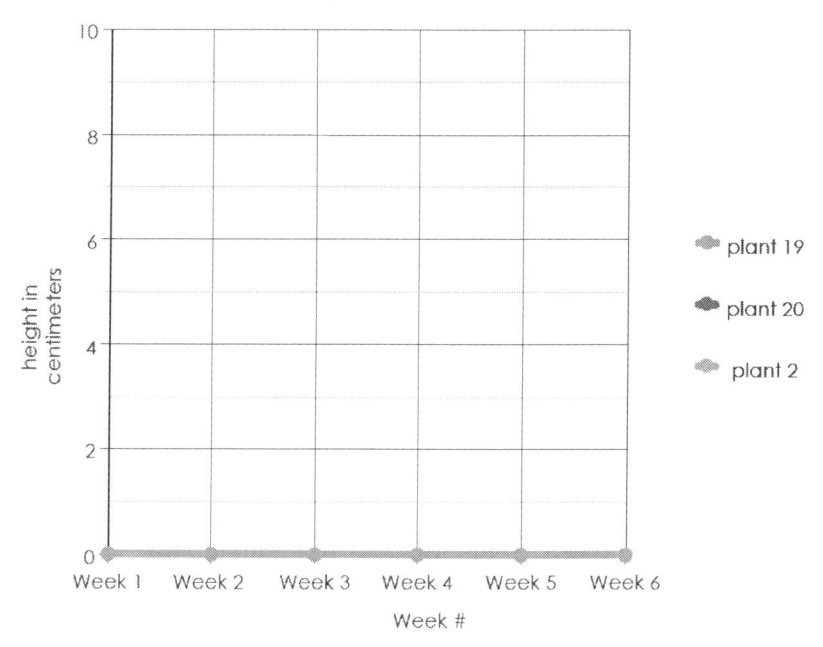

[Graph: height in centimeters vs Week #, showing plant 19, plant 20, plant 2]

Avg. graphs-

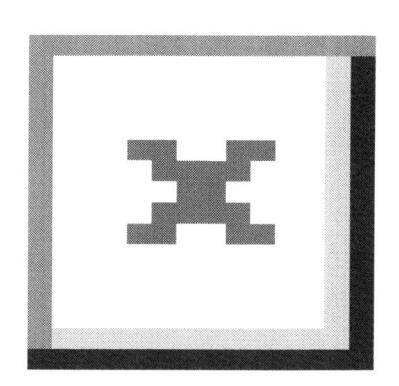

Tallest Blade of Grass (Average between the three plants)

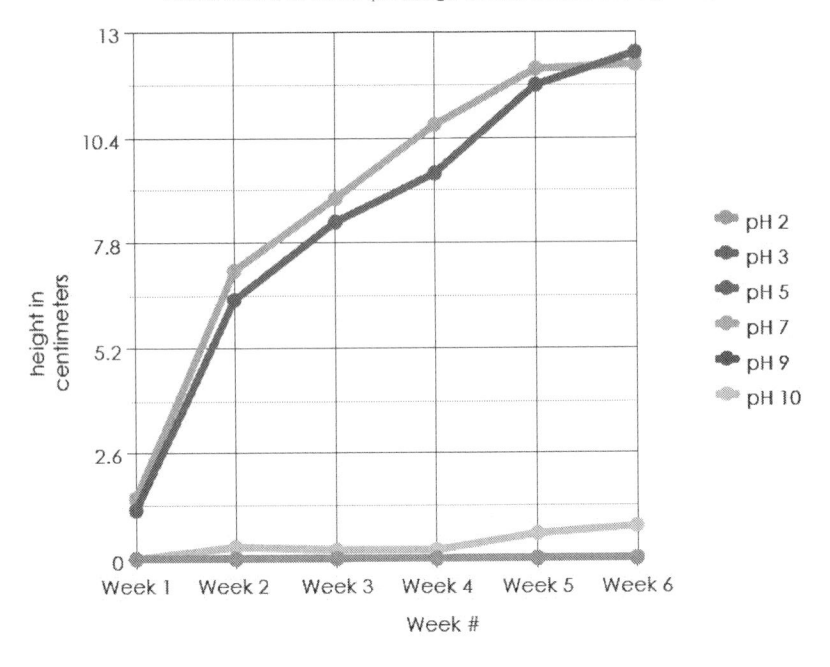

(Line graph showing height in centimeters vs Week # (Week 1 through Week 6) for pH 2, pH 3, pH 5, pH 7, pH 9, pH 10. Y-axis marks: 0, 2.6, 5.2, 7.8, 10.4, 13.)

When analyzing the data we recorded, we observed the amount of blades per cup. The group watered with a pH of 5, had the largest amount of blades. In comparison to the other pH groups that had blade growth it had 102 more blades than the pH of 10 and it had 25 more blades than the pH of 7. In comparing blade height, the pH of 5 also had the tallest blades at the end. Its tallest blade was 0.3 cm taller than the tallest blade that received a pH of 7 and it was 11.7 cm than the blade that received a pH of 10. When comparing color there were not many differences between pH of 5, 7, and 10. The only minor difference was the pH of 10 had 2 brown blades and some weeks the pH groups of 5 and 7 had white tipped blades, with ph of 7 having more. When observing the strength of the blades the pH of 5 was much stronger than the pH of 7 and the pH of 10 never grew enough to be comparable. We did not observe any signs of growth in the other pH groups.

While doing this experiment there may have been slight errors in our experiment. One of these could be that while watering we may not have had exactly the certain pH or given the plant exactly one tablespoon of the solution. Also, another error is when counting the blades for our observations, they were often clumped together and hard to count, which may have led to not precise data. A definite error occurred on December 27, 2010 when the power went out and our plants received about an hour less of light that day.

VI. **Conclusion**:

Our final observations and analysis of the data have led us to a conclusion that the pH of 5 yielded the most plant growth. This conclusion supports our hypothesis that plants receiving a more acidic solution would grow taller. This also proves that overall when giving plants slightly acidic solutions, it will cause it to grow stronger and more efficiently. The purpose of our project being to determine what pH level made plants grow most effectively was met and we clearly found which pH levels not to use when watering plants and which will work best. We concluded that pH levels of: 2, 3, 9, and 12, will show no growth. Also, we concluded that a pH of 10 will show minimal growth. A pH of 5 and 7 will show the best growth, however, a pH of 5 would be the best choice of solution when growing grass.

Application to Society

The results from this experiment will benefit farmers, gardeners, and landscapers. These results could be used in describing the effects of acid rain. Unpolluted rain has a pH value of 5.6 (Acid Rain, 1995), while acid rain pH levels range anywhere below 5.6. When referring to our results, it makes since why all acid rain does not affect the growth of plants. Acid rain with a pH between 5 and 5.6, are not harming the plants, but when analyzing our results they could potentially be increasing their growth. Acid rain with levels below a pH of 5 is probably harmful to the plant growth.

Another use of our conclusions would be for farmers to use it for agricultural advances. Our results showed that the pH of 5 grew more blades of grass. Farmers could use this information if they wanted their crops to grow quicker. Using a solution with some vinegar and water (pH of 5) would allow a farmer to yield more plants than if they used pure water (pH of 7). If farmers were to do this, then it would take less time to grow their crops. Also our results showed that a pH of 5 produced a stronger blade of grass. This could help farmers in several ways. For example, if a farmer used a solution with a pH of 5 to water their plants, overtime they would produce a stronger crop. This could prevent harm that certain rodents or other small animals could cause to the crop.

The results also may be used to create ways to eliminate harmful fertilizers. Normally farmers will use pure water and fertilizers to increase the growth of crops. These fertilizers are harmful to birds and other small animals that may use the crops as their food. If farmers used our watering method (solutions with a pH of 5) instead of fertilizers, it will still increase the growth of crops and eliminate harm to animals.

Most farmers use lime to lower the pH of the soil their crops grow in. Spreading the lime throughout the soil takes time, especially on larger farms. If a farmer were to water the soil with a solution consisting of a pH of 5, they would no longer have to spread the lime. This is because the soil would absorb the solution changing the soil's pH to 5. Watering the soil with a solution of a pH of 5 would be just as effective as spreading lime, but less time consuming.

Landscapers could also benefit from this study. As part of their job, they are required to grow grass. Most landscapers use a method called hydro seeding instead of the normal means of planting grass seed. This is a method in which landscapers spray a mixture of water and grass seed over the certain area that they want grass to grow. If a landscaper were to mix a solution with a pH of 5 with the grass seed instead of pure water, the grass would sprout even quicker. This could help a landscaper get the job done quicker and allow them to get more jobs done per year.

In our experiment we concluded that a slightly acidic solution with a pH of 5 would grow grass more efficiently. This could be helpful in many different ways as we previously mentioned. Our conclusions can help in professional fields such as agriculture and landscaping. This could also help the environment and those animals in it. This information will provide people with a more efficient way to grow plants and crops.

Reference Page:

Acid rain. (1995). http://elibrary.bigchalk.com/elibweb/elib/do/document?urn=urn:bigchalk:US;BCLib;doucument;28013712

Coomes, D A, & Allen, R B (Sept 2007). Effects of size, competition and altitude on tree growth. The Journal of Ecology, 95, 5. P.1084(14). November 01, 2010, Expanded Academic ASAP via Gale: http://find.galegroup.com/gtx/start.do?prodId=EAIM&userGroupName=mlin s somrst

Indoor Gardening Troubleshooting Guide. (2010). http://www.jasons-indoor-guide-to-organic-and-hydroponics-gardening.com/indew.html

Leger, E., Howe, K., Gurevitch, J., Woo, E., Hickman, J., Ashton, I., Lerdau, M. (2007, December 01). The Interaction between Soil Nutrients and Leaf Loss during Early

Establishment in Plant Invasion. Forest Science, (6), 701, http://elibrary.bigchalk.com

Miller, Michelle (2001, April 01). Fulfilling special needs of nurseries. BioCycle, (4), 55, http://elibrary.bigchalk.com

NCDA & CS. Plant nutrients. http://www.ncagr.gov/cyber/kidswrld/plant/nutrient.html

Oram, B. (1986). Ph of common liquids. http://www.water-research.net/ph.htm

Pimentel, D., Berger, B., Filiberto, D., Newton, M., al, E. (2004, October 01). Water Resources: Agricultural and Environmental Issues. Bioscience, (10), 909, http://elibrary.bigchalk.com

Plant pH preference range lists by category. (2007). http://homeharvest.com/plantphpreference.html

Plants and ph. (2004). http://angel-patti03.tripod.com/id8.html

Soil gardening ph. (2010). http://www.jasons-indoor-guide-to-organic-and-hydroponics-gardening.com/

The Effect of Rising Ph Levels on Foraminifera
By Tyler Barron

Summary:

In ninth grade I conducted a project at the Woods Hole Oceanographic Institution on the effect of rising pH levels due to ocean acidification on the viability of foraminifera. That year I was unable to directly adjust the CO_2 levels, which are the cause of rising pH levels in the ocean. However, when science fair began again, I talked to one of the scientists I worked with in 9th grade to see if they would mentor for a project. I eventually ended up working on this experiment as WHOI with Dan McCorkle testing various CO_2 levels on the growth of bay scallops. Bay scallops were chosen for this experiment because they are calcifying organisms. They represent an organism that has the potential to be negatively impacted by rising CO_2 levels in the ocean. Based on the results of this experiment, implications of increasing CO_2 levels due to anthropogenic and natural inputs of nitrogen and CO_2, could be inferred. Increasing amounts of nitrogen accelerate decomposition in eutrophic waters, which increase CO_2 levels in estuary and bay waters. Deforestation, cement production and the burning of fossil fuels also contribute to the increasing amount of CO_2 in the atmosphere, and when absorbed by the world's ocean lower the carbonate ion saturation. An understanding of how bay scallops react to increased pCO_2 levels may help to indicate the extent in which marine environments, and economies dependant upon these marine resources, will be affected.

Abstract:

Since the Industrial Revolution, atmospheric CO_2 levels have risen from 280 to 391 ppm; 10 billion tons of CO_2 are released annually. When released into the atmosphere though natural processes or anthropogenic processes, CO_2 is absorbed by the world's oceans, leading to a net decrease in carbonate ions (CO_3^{2-}). Carbonate ions are an essential component to marine chemistry because they are required for the production of $CaCO_3$, which aids calcareous organisms in the process of calcification. Although the oceans are currently super-saturated with aragonite and calcite, forms of $CaCO_3$, models indicate that lysoclines will decrease with rising atmospheric CO_2 levels. In addition to ocean acidification, natural and anthropogenic inputs of nitrogen have also contributed to increasing CO_2 levels. Algal blooms that develop as a result of eutrophication sink to the bottom of bays, ponds and estuaries. The subsequent decomposition of this algae increases CO_2 levels and decreases CO_3^{2-} concentrations. In estuary and bay environments with naturally eutrophic waters, calcareous organisms, such as *Argopecten irradians*, will be forced to either allocate more energy to calcification or stored energy reserves. If organisms are unable to adapt, ecosystems and marine fisheries will begin to suffer.

This project modeled CO_2 levels in estuary and bay environments, in particular Waquoit Bay, East Falmouth,

MA, where the concentration of CO_2 at certain locations exceeds 2000 ppm. Bay scallops were exposed to environments containing 380, 840, 1300 and 2000 ppm CO_2 for 29 days in order to determine if elevated CO_2 levels had an effect on their growth. It was hypothesized that scallop growth would decrease as CO_2 increased, however, this hypothesis was partially supported because the data showed that scallops mass remained relatively constant through 380 to 1300 ppm, but dropped off at a threshold of 2000 ppm.

Introduction:

Air bubbles trapped in ice sheets provide a record of the past atmospheric carbon dioxide, from which ocean pH can be calculated. Long cores extracted from the ice show that over the last 800,000 years until the mid-1800s, atmospheric CO_2 had never exceeded 280 ppm, or a pH of 8.2 (Ocean Acidification: Questions Answered, 2009). However, since the beginning of the Industrial Revolution in the 1830s, the burning of fossil fuels, deforestation and cement production has released more than 440 million metric tons of CO_2 into the atmosphere, leading to the current atmospheric CO_2 concentration of 391 ppm. The increased levels of atmospheric carbon dioxide have continuously been absorbed by the ocean in a process that contributes to ocean acidification, the decrease in pH of the world's oceans. In this process, atmospheric CO_2 is absorbed to form aqueous CO_2. The aqueous CO_2 reacts with seawater (H_2O), producing carbonic acid (H_2CO_3). Carbonic acid rapidly dissociates to hydrogen ions (H$^+$), resulting in ocean acidification, and bicarbonate ions (HCO$_3^-$). The H$^+$ combine with carbonate ions (CO$_3^{2-}$), resulting in a net decrease in carbonate ion concentration. This process is magnified by the 10 billion tons of CO$_2$ that are released annually, 2 billion of which are absorbed by the oceans; in effect, everyday each American adds 40 lbs of CO$_2$ to the oceans (Doney et al., 2009). Since the beginning of the industrial revolution, rising atmospheric CO$_2$ levels have driven 118±19 petagrams of carbon into the ocean (Glenhill et al., 2009), primarily into the upper 1000 meters. The subsequent 0.1 drop in pH, although seemingly small, represents a 30% rise in acidity due to the logarithmic measure of the pH scale. Atmospheric CO$_2$ levels are expected to reach 500 ppm by the middle of the century, and 800 ppm by the end, resulting in a 0.3 drop in pH or a 150% increase in hydrogen ions (Feely et al., 2009). This increase of hydrogen ions has not only noticeably decreased pH levels, but carbonate ion concentrations as well. Carbonate ions are an essential part to marine chemistry because they are required for calcification, the process in which organisms use dissolved carbonate ions to produce CaCO$_3$ and, as a result, calcareous shells and skeletons.

Although the oceans are currently supersaturated with calcium carbonate (CaCO$_3$), the decline in CO$_3^{2-}$ has had a corresponding effect on the degree to which seawater is saturated with carbonate mineral phases, nominally aragonite and calcite. At equilibrium, the product of the calcium concentration and the carbonate concentration is equal to a constant called the apparent solubility product. This apparent solubility product differs with temperature, salinity, and pressure, and differs among calcium carbonate minerals (Feely et al., 2009). Under-saturation for aragonite occurs for most open ocean surface waters when the carbonate ion concentration drops below 66 µmol kg^{-1}. Model projects indicate that aragonite under-saturation will start to occur by about 2020 in the Arctic Ocean and 2050 in the Southern Ocean (Feely et al., 2009). Calcite under-saturation occurs when carbonate ion concentration drop below 42 µmol kg^{-1}. By 2095, most of the Arctic Ocean and some part of the Bering and Chukchi seas will be under-saturated with respect to calcite (Feely et al., 2009). The differing saturation horizons, or lysoclines, of each mineral indicate that aragonite is much more soluble that calcite. Consequentially, organisms that use aragonite may be more vulnerable than those that use calcite to produce tests or skeletons. Shells and skeletons of unprotected $CaCO_3$ begin to dissolve when degrees of their component minerals drop below $\Omega = 1$ and because many organisms have optimal calcium carbonate precipitation rates at supersaturated states, decreases in CO_3^{2-} concentrations could decrease calcification rates for a number of those species. Although most surface waters are supersaturated with respect to $CaCO_3$,[Ω (saturation state) values for aragonite: 2-4; Ω values for calcite: 4-6] which means calcification is favored over dissolution (Feely et al., 2009), increases of CO_2 and lowering of pH levels raise the lysoclines for both minerals closer to the surface. Increases in anthropogenic carbon dioxide over the past 250 years have caused aragonite and calcite saturation states to become close to the surface by 50m to 200m depending on the location (Feely et al., 2009).

Fifty-five million years ago in the Cenozoic Era (the Palaeocene-Eocene Thermal Maximum (PETM)), the earth experienced a rise in temperature (11°F) over a period of 10,000 years. Atmospheric CO_2 levels rose, cause the depth of calcium carbonate saturated water to become much shallower, which eventually led to the demise of many marine calcifiers. On large timescales (>10,000 years) there is a natural balance maintained between the uptake and release of CO_2 on earth by the millennial-scale mixing cycles where surfaces waters circulate through the deep ocean and back, resulting in the dissolution of carbonate sediments on the sea floor into carbonate ions which helped to neutralize some of the acidity, but on shorter timescales (>1,000 years) the ocean is unable to compensate for the increasing carbon dioxide. As a result of the drastic increase of CO_2, about 50% remains in the upper 10% of the ocean (Ocean Acidification: Questions Answered, 2009). The ability of sediments to regulate ocean chemistry and neutralize the acidity is simply too slow- taking place over more than 1,000 years- and so ocean pH and the amount of available carbonate ions is now declining (Ocean Acidification: Questions Answered, 2009). The rate of acidification today is 10 times faster than anything experienced since the death of the dinosaurs 65 million years ago.

In contrast to the calcium carbonate in the form of dead shells, which at shallow water depth is buried in the sediment and trapped for a long time, the $CaCO_3$ of shells that sink in deep water depth get dissolved, thereby releasing the carbon. However, the current increased rate of atmospheric CO_2 being absorbed by the oceans is resulting in an imbalance in the carbonate compensation depth (CCD), the depth at which all carbonate is dissolved. As the pH of the ocean falls this results in a shallowing of the lysocline and the CCD, thus exposing more of the shells trapped in sediments to under-saturated conditions causing them to dissolve, which helps to further buffer ocean acidification but over a long time scale of a thousand years (Ocean Acidification: Questions Answered, 2009). For many marine organisms, their larval stages, juveniles and adults may be affected differently by ocean acidification. Studies on larval marine bivalves showed that they are expected to be more vulnerable than adults to increased CO_2 because they biomineralize aragonite, the more soluble form of $CaCO_3$ rather than calcite, the predominant material used in adult shells (Talmage et al., 2009). As aragonite becomes less saturated, the energetic costs for biomineralization will become progressively greater, forcing calcifying organisms to either allocate more energy to calcification at the cost of growth and stored energy reserves, or to accept a lesser degree of calcification in exchange for greater growth and stored energy (Talmage et al., 2009). Due to the fact that metamorphosis is energetically demanding and relies entirely on energy stores accumulated during the larval stage, it may lead to the reduced survival rate of young marine organisms. Slowed growth and calcification causes prolonged time spent in the water column (column of water from surface to bottom sediments), which will result in increased mortality rates from predation and disease, both of which are common in the water column. Although most species possess some capacity to acclimate, adaptive physiological tolerance of larvae may not be sufficient to sustain populations of calcifying organisms. The rate at which ocean acidification is occurring may be too fast for organisms to adapt and the resulting stress may possibly affect certain species' ability to compete in their environment, thereby limiting economically important fisheries.

Declines in calcifying organisms' populations, due to ocean acidification, may result in increased competition among predatory finfish, affecting the overall abundance of some commercial finfish and species diversity. Current estimates suggest that worldwide capture fisheries and aquaculture provided 143.7 million metric tons, valued at US$246 billion, of marine products in 2006; of the global total, 110 million metric tons were used directly as food for humans. The seafood produced by marine capture-fisheries and aquaculturists provided an average of 13.6 kg per capita of food (Cooley et al., 2009). If ocean acidification reduces calficiers' harvests, there could be a decrease by millions of metric tons and billions of dollars in revenues, resulting in additional costs for seafood consumers, aquaculturists and fishermen. Loss of calcifying organisms, which support marine ecosystems, and the resulting ecosystem services that generate income, could substantially affect global economies. Some of the most commercially important ecosystems that will be affected by ocean acidification are estuary environments and coastal waters.

Estuaries, which are a partially enclosed bodies of water, and their surrounding coastal habitats where saltwater from the ocean mixes with fresh water from rivers or streams, offer habitats for more than 75 percent of the U.S. commercial fish catch and an even higher percentage of the recreational fish catch. The total fish catch in estuaries contributes $4.3 billion a year to the U.S. economy (NOAA Ocean Service, 2008, Internet). Coastal estuaries and biomes are among the most biologically productive ecosystem services; they provide commercial and recreational fisheries, invertebrate nursery grounds, water purification, flood and storm surge protection and human recreation (Talmage et al., 2009). Because they are generally shallower and lack carbonate sediment deposits that aid in the reversal of acidification, coastal marine habitats are more susceptible to changes in pH and CO_3^{2-} concentration. Low alkalinity of freshwater inflow in addition to the slow rate of deep ocean circulation both contribute to an overall decrease in pH and a subsequent decrease in the saturation state of calcium carbonate (Feely et al., 2009).

As ocean acidification diminishes the lysocline each year, natural upwelling events more often cause under-saturated water, as well as nutrient-rich water to well up and flow to the shore. Coastal and estuarine organisms that form calcium carbonate shells are unaccustomed to such events, and periodic exposure to these significantly different conditions may adversely affect these ecosystems (Ocean Acidification: Questions Answered, 2009). Nutrients that make their way into coastal waters through deep-sea upwelling increase algal activity, and, through their decomposition, increase carbon dioxide levels. In addition to upwelling, eutrophication, the process in which excessive nutrients are released into a body of water, thereby promoting biological activity, contributes to an overall decrease in carbonate ion concentration through the increase in aqueous CO_2. Estuary waters tend to be naturally eutrophic due to the concentration of land-derived nutrients in confined channels. These nutrient inputs, most commonly resulting from wastewater, fertilization and industrial site pollution, lead to large algal blooms (Talmage et al., 2009). When these blooms collapse and sink to the seabed, the subsequent respiration of bacteria decomposing the algae leads to an increase in CO_2 and decline in pH (Ocean Acidification:

Questions Answered, 2009). These increasing carbon dioxide levels decrease the carbonate ion concentration of the water, making it exceedingly difficult to calcifying organisms to produce shells and skeletons. As important estuary and coastal water ecosystems become under-saturated in calcium carbonate, organisms will be forced to adapt to these changing conditions. Consequentially, food chains will be altered and affecting certain species' ability to compete in their environment; among these species are the bay scallops. *Argopecten irradians* are bivalve mollusks that use carbonate ions in the process of calcification to form their shells. They are commonly found in the North Atlantic in bay or estuary environments. Bay scallops are epibenthic in nature, meaning that that live on the surface of the bottom rather than in the sediments like many other bivalves (The Ecology of the Waquoit Bay Estuary, 1996). They attach themselves to eelgrass or the seabed, anchoring themselves with their foot. They swim by exerting a water jet through the rapid contraction of their valves. By exercising the same valves, they are able to bury themselves into the sediment to avoid strong currents. In their first stages of development, the scallops' embryos and larvae float freely in the water column. After the mass of the scallops' developing valves overwhelms its swimming capabilities, it settles to the bottom. Shell deposition slows in the winter but gradually increases as summer nears. Maturity for bay scallops is reached at the end of their first year, after which growth is directed to shell development (The Ecology of the Waquoit

Bay Estuary, 1996). However, in high CO_2 environments, *Argopecten irradians* are put under greater stress to produce calcium carbonate shells. As ocean acidification progresses due to increased anthropogenic carbon dioxide levels, estuary habitats with already elevated CO_2 concentrations will exacerbate conditions for bay scallops.

Ocean acidification posses a threat to the survival and continuation of many calcifying species, and as a result, the ecosystems in which they live. Ecosystem services dependant upon these marine environments will also be affected by the ramifications of rising pH levels. In particular, estuarine and coastal water habitats will be most greatly impacted by anthropogenic increases in atmospheric CO_2 due to the upwelling of nutrients and under-saturated water, in addition to the eutrophic decomposition of algae that already takes place. Organisms such as *Argopecten irradians*, are and will be forced to acclimate to altered concentrations of CO_3^{2-}. The subsequent inhibition of larval metamorphosis and reproduction will lengthen their time in the water column and force them either to allocate more energy to calcification or growth and stored energy supplies.

With an increase in CO_2 of 2 ppm per year due to the burning of fossil fuels, deforestation and cement production (Ocean Acidification: Questions Answered, 2009), in addition to the anthropogenic and natural input of nutrients, which, through the process of eutrophication and decomposition, decreases carbonate ion saturation, estuary and coastal water environment, and the calcifying organisms that inhabit and support these essential ecosystems, will be irreversibly altered.

Methodology:

Bay scallops collected from Waquoit Bay, which had been placed into a holding bin with similar conditions to their previous environment were transferred to the lab environment where the individual scallops were cleaned with a toothbrush to remove any debris. Since three scallops were needed for each bucket, tally marks were made on each animal to indicate whether the scallop was a one, two or three. After the scallops were cleaned and labeled they were moved into a new bin adjacent to the initial holding bin.

The following day, the numbered scallops were taken from their holding bin, placed into a bucket filled with seawater and brought into the lab. Three scallops, labeled one, two and three, were placed into a cup containing seawater. This cup was then emptied into a massing apparatus to try to ensure the scallops would not take up any air. The massing apparatus consisted of a clear plastic box with a balance placed on top. Descending from the balance into the water was a small platform on which the organisms were placed. Scallops labeled either one, two or three were placed onto the platform in order to determine a buoyant mass measurement. The mass of the scallop, in grams, was recorded after the balance stabilized or when the mass began to only change within the ten-thousandths place. Scallops two and three were also measured using this same process. Eight tanks, 380 A, 380 B, 840 A, 840 B, 1300 A, 1300 B, 2000 A and 2000 B, which had previously been set up by Meredith White, were configured to bubble at four different pCO_2 levels. Three scallops were placed into each of these tanks, however, due to time constraints, only 2 scallops were massed for tanks 840 A, 840 B, 1300 B and 2000 B. The remaining labeled scallops were also placed into their respective tanks, but were not massed.

In order to feed the scallops, a cell count of an algae solution was first taken. Using a pipette, a sample of the solution was placed onto a slide. With a grid on the slide, the algae were viewed under a microscope to determine the number of algal cells in a given area. Due to the fact that the area (0.25 mm²) and depth (0.1 mm) of the grid was known, the number of cells in a given volume was calculated. The number of mL of algal culture that would be given to each tank was calculated using the equation: ((grams dry weight/adult scallop)* (% dry weight ration/100%)*($1*10^6$ cells/0.0002 g)* # of adult scallops*(1ml/# of algae cells)).

10 L of the same algae solution, primarily containing *Tetraselmis,* was added to a bucket. Based on the previous calculation, it was determined that 322 mL of algal culture would be added to each tank. Using a beaker, 322 mL of the solution was then added to each tank. This process was

repeated every few days. However, due to the fact that it was too difficult to grow the algae within the time frame of this experiment, a faster and easier alternative was found to feed the scallops. This alternative was Instant Algae. 30 mL of the Instant Algae was mixed into 600 mL of water. Based on the set number of cells per volume in the Instant Algae, it was determined that 24 mL of the newly created solution would be added to each tank. With a syringe, the algae was taken from the 630 mL solution and administered to each tank. This process was also repeated every few days until the end of the experiment.

The scallops were taken from their tanks 29 days after the start of the experiment in order to be massed. Scallops labeled one, two and three were taken from each tank and placed into Ziploc™ containers containing a solution of $MgCl_2$. $MgCl_2$ was used to anaesthetize the scallops and stop them from moving on the balance. It was also used to open up the scallops so that they would release any air bubbles trapped inside. The $MgCl_2$ was not originally used to mass the scallops because no scallop anesthesia was known at the time. After the scallops sat in the solution for roughly 3-4 minutes, they opened up and were placed onto the platform of the balance for a buoyant mass measurement. When the balance stabilized, the mass was recorded and the scallop was taken out of the massing apparatus and placed into a Ziploc™ bag. The balance was then tared and the remaining two scallops were massed and bagged. This entire process was repeated for all three scallops from each tank until all of the scallops had been massed. The scallops were then placed into a freezer where they stayed until it was time to cut and measure them.

To clean the scallops, they were taken from their bags, and with a set of forceps, their entrails were removed. Then, using a caliper, the length, hinge length and extension length was recorded for each scallop. After the scallops were measured, they were taken to the IsoMet 1000 Precision Saw to be but in half vertically. The measured and cut scallops were then placed into small metals tins, which had been labeled on the bottom to identify the tank and number of the scallop.

The following day, two scallops from each pCO_2 level were placed under a microscope in order to take a picture to determine their growth. The scallops were placed onto a wax container to be stabilized. Using the program SPOT Advanced with a microscope attached to the computer, pictures of the two scallops were taken at 0.75x magnification. Lengths of the extension of growth, if visible, were calculated using a measurement device in the program.

Data:

Table 1. Initial Mass, Final Mass, Mass Differences and Percent Of Growth in Each Treatment

Treatment (ppm CO_2)	Initial Mass (g)	Final Mass (g)	Difference	% Of Growth
380	2.97	2.89	-0.08	7.98
	2.56	2.73	0.17	6.72
	2.38	2.40	0.02	0.83
	2.53	2.50	-0.03	-0.03
	3.63	3.84	0.21	5.75
	2.72	2.95	0.22	8.18
	4.49	4.69	0.20	4.44
	3.63	3.65	0.02	0.45
	2.44	2.62	0.18	7.59
840	2.66	2.81	0.15	5.71
	2.55	2.66	0.12	4.54
	3.17	3.23	0.06	1.93
	3.00	3.11	0.11	3.80
	2.77	2.84	0.07	2.44
	2.81	2.95	0.14	5.15
1300	3.47	3.68	0.22	6.24
	2.17	2.17	0.00	0.18
	4.99	5.13	0.15	2.98
	2.99	3.77	0.78	26.11
	1.61	1.77	0.15	9.45
2000	2.61	2.61	0.01	0.26
	2.65	2.69	0.04	1.55
	3.18	3.20	0.02	0.56
	2.46	2.39	-0.07	-2.84
	1.48	2.02	0.55	36.95

Table 1. Initial Mass, Final Mass, Mass Differences and Percent Of Growth in Each Treatment

Treatment (ppm CO_2)	Initial Mass (g)	Final Mass (g)	Difference	% Of Growth
380	2.97	2.89	-0.08	7.98
	2.56	2.73	0.17	6.72
	2.38	2.40	0.02	0.83
	2.53	2.50	-0.03	-0.03
	3.63	3.84	0.21	5.75
	2.72	2.95	0.22	8.18
	4.49	4.69	0.20	4.44
	3.63	3.65	0.02	0.45
	2.44	2.62	0.18	7.59
840	2.66	2.81	0.15	5.71
	2.55	2.66	0.12	4.54
	3.17	3.23	0.06	1.93
	3.00	3.11	0.11	3.80
	2.77	2.84	0.07	2.44
	2.81	2.95	0.14	5.15
1300	3.47	3.68	0.22	6.24
	2.17	2.17	0.00	0.18
	4.99	5.13	0.15	2.98
	2.99	3.77	0.78	26.11
	1.61	1.77	0.15	9.45
2000	2.61	2.61	0.01	0.26
	2.65	2.69	0.04	1.55
	3.18	3.20	0.02	0.56
	2.46	2.39	-0.07	-2.84
	1.48	2.02	0.55	36.95

Figure 1. Percent Increase in Mass (g) vs. Aragonite Saturation

Table 1. Initial Mass, Final Mass, Mass Differences and Percent Of Growth in Each Treatment

Treatment (ppm CO_2)	Initial Mass (g)	Final Mass (g)	Difference	% Of Growth
380	2.97	2.89	-0.08	7.98
	2.56	2.73	0.17	6.72
	2.38	2.40	0.02	0.83
	2.53	2.50	-0.03	-0.03
	3.63	3.84	0.21	5.75
	2.72	2.95	0.22	8.18
	4.49	4.69	0.20	4.44
	3.63	3.65	0.02	0.45
	2.44	2.62	0.18	7.59
840	2.66	2.81	0.15	5.71
	2.55	2.66	0.12	4.54
	3.17	3.23	0.06	1.93
	3.00	3.11	0.11	3.80
	2.77	2.84	0.07	2.44
	2.81	2.95	0.14	5.15
1300	3.47	3.68	0.22	6.24
	2.17	2.17	0.00	0.18
	4.99	5.13	0.15	2.98
	2.99	3.77	0.78	26.11
	1.61	1.77	0.15	9.45
2000	2.61	2.61	0.01	0.26
	2.65	2.69	0.04	1.55
	3.18	3.20	0.02	0.56
	2.46	2.39	-0.07	-2.84
	1.48	2.02	0.55	36.95

Table 2. Total Lengths, Lengths Added, % Extension and Hinge Lengths Of Scallops in Each Treatment (Measurements taken with caliper)

Treatment ppm CO_2	Hinge Length (cm)	Total Length (cm)	Length Added (cm)	% Of Extension
380	2.13	3.62	0.38	11.73
	2.62	3.61	0.00	0.00
	2.37	3.55	0.00	0.00
	2.83	3.36	0.12	3.70
	1.96	3.57	0.90	33.71
	2.33	3.96	0.44	12.50
840	2.43	3.51	0.50	16.61
	2.31	3.31	0.22	7.12
	2.52	3.92	0.57	17.01
	2.52	3.48	0.19	5.78
	2.39	3.42	0.38	12.50
	2.40	3.71	0.21	6.00
130	2.34	3.88	0.50	14.79
	2.11	3.10	0.11	3.68
	2.92	4.21	0.35	9.07
	2.70	4.08	0.37	9.97
	2.09	2.61	0.21	8.75

Discussion:

The results from this experiment suggest that CO_2 levels at a high enough level decrease the growth of bay scallops. Figure 1 shows percent increase in mass vs. aragonite saturation state corresponding to each CO_2 level, 1.61, 1.09, 0.74 and 0.51 for 380, 840, 1300 and 2000 ppm, respectively. A threshold for growth was found 2000 ppm, where subsequent growth was inhibited. Initial masses of scallops placed in the various CO_2 environments ranged from 2.38-4.49 g, 2.55-3.17 g, 1.61-4.99 g and 1.48-3.18 g while final masses ranged from 2.40-4.69 g, 2.66-3.23 g, 1.77-5.13 g and 2.02-3.20 g for 380, 840, 1300 and 2000 ppm, respectively (Table 1). Figure 2, which plots percent of shell extension vs. aragonite saturation state corresponding to each CO_2 level, shows that the scallops in each treatment extended in length, although in some cases, they had lost mass. Percent of shell extension ranged from 0.00-33.71 for 380 ppm, 5.78-17.01 for 840 ppm, 3.68-14.79 for 1300 ppm and 5.79-9.73 for 2000 ppm (Table 2).

An interesting observation that was made was that although the length of the scallops increased, the mass increase showed little or even negative growth for the 2000-ppm treatment, indicating that although the scallops lost mass, their shells continued to grow. This agrees with the results of a previous experiment done by Dan McCorkle[1], which also

---

[1] Dan McCorkle, Associate Scientist, Department of Geology and Geophysics,

observed that when scallops were subjected to elevated CO_2 levels they lost mass, but gained length. This could be due to an evolutionary trait, in which the scallops lose mass from their inner or outer shell in order to increase their length. However, the loss of mass from the scallop shell could also be due to involuntary decalcification by the low carbonate ion saturation of the water.

These are a few sources of error in this experiment that could have altered the results. The first possible, and likely, source of error was in the initial measurements of the scallops. Due to the fact that they were not anesthetized in a solution of $MgCl_2$, like they were for the final measurements, there was the possibility of air bubbles being trapped inside their shells. If this were the case, as was suspected for 1300 B1 and 2000 B2, the initial mass measurements would have been lower, increasing the final mass measure, thereby leading to a higher percent increase of mass. Another possible source of error may have stemmed from the feeding regiments of the scallops. Since the scallops were fed live algae for the beginning of the experiment, and then switched to a solution of Instant Algae, this could have negatively impacted their mass or extension.

The results of this experiment suggest that environments with CO_2 levels at or above 2000 ppm, such as those found in

Waquoit Bay, will have adverse effects on scallop growth. These results bring about concern not only scallops, but also about other calcifying organisms such as clams, mussels and snails. Their decrease in viability could lead to subsequent increase in predation by finfish and other predators. This increased predation could alter food chains, negatively impacting the estuarine habitat and its important role in producing food for humans.

There are many ways this project could be expanded upon. If time allowed, more replicates and tanks would have been used. The various CO_2 levels could have been tested on juvenile and adult scallops to see if there was a difference in mass increase or extension between the two. The scallops could also have been subjected to the CO_2 treatments for various time periods to see if short or long-term exposure had the same effect. Also, a larger variety of pCO_2 levels could have been used to see exactly when the threshold limit occurs for mass growth. Lastly, increased CO_2 levels could have been tested on different organisms to see which organisms were most tolerant to these changes, and to ultimately infer if increased CO_2 levels due to anthropogenic processes will adversely affect their growth and viability.

The results of this experiment suggest that high pCO_2 concentrations, at or above 2000 ppm, will have adverse effects of scallop growth. However, this data also helps to infer more general consequences of elevated CO_2 levels. In environments that already experience high CO_2 levels, which included Child's River and other sections of Waquoit Bay, calcareous organisms may be struggling to effectively grow and survive. This disadvantage could effect, and may already be affecting this estuary ecosystem. Local scallop harvests have gradually been decreasing in Waquoit Bay from 200,000 L/yr in 1965 to 20,000 L/yr in 1995 (Waquoit Bay Watershed Ecological Risk Assessment: The effect, of land-derived nitrogen loads on estuarine eutrophication, U.S. Environmental Protection Agency). This decrease in scallop harvest could be due to the ever increasing inputs of anthropogenic CO_2 into the atmosphere, as well as the increasing anthropogenic and natural inputs of nitrogen into the watershed and estuary, and subsequent increase in CO_2 due to the bacterial decomposition of algal blooms. This diminishing scallop population also poses a threat to the possible remediation of Waquoit Bay. Scallops and other bivalves which help filter water through algal ingestion serve as a mechanism to remediate the eutrophication caused by natural and anthropogenic processes.

This same eutrophication process, due to an increase of nitrogen, which ultimately leads to increased CO_2 levels, negatively effects eelgrass populations. Due to the fact that eelgrass increases the availability of particulate food, therefore increasing growth rates, it serves as a nursery ground for juvenile bay scallops and a protected habitat for adult bay scallops, the highest densities of these bivalves have been found to live in eelgrass meadows. Juvenile calcifying organisms, due to eutrophication and low carbonate ion concentration will be faced with either a loss of calcification or a loss of growth and stored energy reserves. Other estuaries and bays, with shallow water columns and naturally or artificially eutrophic waters, could also be experiencing high CO_2 conditions. If the atmospheric and eutrophic addition of CO_2 is to increase, these CO_2 concentration in parts of these waters that are naturally high will further increase, and the CO_2 concentrations of surrounding waters will increase as well. In addition to important estuary and bay ecosystems being affected by these changing CO_2 levels, deep-sea environments could begin to be impacted also. Carbonate ions saturations states, could decrease to a point in which they are detrimental to calcareous organisms, forcing them to adapt to these changes. As a result, predation among finfish will increase, altering delicate food webs and decreasing species diversity. Both estuary and deep-sea ecosystems, which are vital to commercial fisheries, may begin to suffer, subsequently impacting economies and businesses that are dependant upon marine resources. If ocean acidification and high inputs of nitrogen are to increase, calcifying organisms and the food chains that rely on them will begin or continue to alter (Talmage et al., 2009).

The anthropogenic factors that contribute to the overall increase of pCO_2 and decrease in carbonate ion concentrations are projected to escalate over the next century, and if they continued unregulated by local and national governments, marine ecosystems will be irreversibly changed. Although the Massachusetts Title 5 Septic Inspection & Testing Laws was passed in order to regulate watershed pollution, larger organizations must be created to research and prevent the eutrophication of fresh and saltwater environments. Organizations such as the Waquoit Bay National Estuarine Research Reserve and the Federal Ocean Acidification Research and Monitoring have been established to discover more about the effects of increasing CO_2 levels on the environment but greater individual participation will be required to reduce the 40 Lbs of CO_2 added to the oceans by each American daily. Without full cooperation to reduce to the causes of ocean acidification and eutrophication, marine ecosystems and global economies could be irrevocably altered.

Conclusion:

The purpose of this experiment was to test of affect of various pCO_2 levels on the growth of *Argopecten irradians*. Data collected from this experiment indicated a threshold limit for percent of mass increase at 2000 ppm. All other CO_2 levels maintained a constant percent that averaged between 3.93 and 4.71 percent, while the scallops in 2000 ppm had an average growth of -0.12 percent. The percent of extension (with caliper) ranged from 7.51 to 10.84, for 380 to 2000 ppm, suggesting that although certain scallops had been exposed to high pCO_2 levels with the possibility of decalcification due to an under saturated carbonate ion concentration, they still lost mass. This data gives evidence to the possible implications of elevated pCO_2 levels as a result of anthropogenic processes that add to ocean acidification and eutrophication. It also suggests the possibility of an advantageous trait within scallops, which cause them to continue shell extension at the expense of a loss of mass growth. It was hypothesized that scallop growth would decrease at pCO_2 levels increase. This hypothesis was partially supported because although there was a decrease in shell mass at 2000 ppm, shell extension was present in all CO_2 levels, and treatments between 380 and 840 ppm experienced similar increase in mass.

Bibliography:
Cohen, Anne L., Michael Holcolm. 2009. Why corals care about ocean acidification:
Uncovering the mechanism. Oceanography. Volume 22.
Cooley, Sarah, Hauke Kite-Powell, Scott Doney. 2009. Ocean acidification's potential
to alter global marine ecosystem services. Oceanography. Volume 22.
Doney, Scott, William Balch, Victoria Fabry, Richard Feely. 2009. Ocean acidification:
A critical emerging problem for the ocean sciences. Oceanography. Volume 22.

EPOCA. 2009. Ocean Acidification: Queestions Answered

Feely, Richard, Scott Doney, Sarah Cooley. 2009. Ocean Acidification: Present conditions and future changes in a high CO_2 world. Oceanography. Volume 22.

Glenhill, Dwight, Rik Wanninkhof, Mark C Eakin. 2009. Observing ocean acidification from space. Oceanography. Volume 22.

Kolbert, Elizabeth. 2006. The Darkening Sea: What carbon emissions are doing to the ocean. The New Yorker.

Miller, Whitman A, Amanda Reynolds, Cristina Sobrino, Gerhardt Riedel. 2009. Shellfish face uncertain future in high CO_2 world: Influence of acidification on oyster larvae and calcification and growth in estuaries. PLoS ONE.

NOAA Ocean Service Education. 2009. "Estuaries." <www.oceanservice.noaa.gov/education/kits/estuaries/estuaries02 economy.html>

NOAA Office of Coastal Resource Management, Sanctuaries and Reserve Division. 1996. The Ecology of the Waquoit Bay National Estuarine Research Reserve. NOAA.

Talmage, Stephanie, Christopher Gobler. 2009. Effects of past, present, and future ocean carbon dioxide concentrations on the growth and survival of larval shell fish. PNAS.

U.S. Environmental Protection Agency. 2002. "Waquoit Bay

watershed

Ecological risk Assessment: The effect of land derived nitrogen loads on estuarine eutrophication"

<www.oaspub.epa.gov.org/ncea/cfm/recordisplay.cfm?deid=15221>

iHeat: A Personal Thermal to Electric Energy Conversion Apparatus

By Curtis Belmonte

Abstract

A prototype thermoelectric harvesting device was developed for metabolic waste heat recovery. The apparatus was designed to incorporate a bismuth telluride thermoelectric generator in order to utilize the Seebeck Effect to convert excess heat produced in the process of metabolic heat generation into low-power, DC electric output energy. The efficiency of the device was also improved by the incorporation of a liquid heat exchanger system designed to maximize the temperature differential across thermal junctions. Additionally, the prototype was designed to accommodate an array of thermoelectric generators and a DC-to-DC step-up conversion circuit comprising a series of charge pumps in order to increase its voltage output to a level sufficient to supply a charge to a standard lithium-ion video iPod battery. Testing was conducted to determine a baseline epidermal surface temperature, and the voltage and current output of the thermoelectric generator were measured using a rubber hot water bottle of various surface temperatures as a controlled heat source. The final prototype design was constructed and affixed to an article of polar fleece headgear in order to optimize its efficiency and practicality as green technology and a viable alternative source of energy.

Introduction

The constant and ever-increasing worldwide demand for energy has led to renewed research into alternative sources of energy. Many large-scale efforts to utilize unharnessed sources of energy are underway, but the engineering of a device to provide immediate and efficient clean energy on a small and marketable scale is an engineering goal that has yet to be adequately accomplished (Angrist, 1982). Furthermore, many of the existing methods for generating or utilizing various forms of energy lack efficiency or are otherwise impractical. Rather than producing the ideal maximum output energy, many systems lose much of the input energy in transfer. Most commonly, this lost energy is expelled as excess, waste heat. However, this waste heat has the potential to be utilized as a viable source of energy. Because the body also produces waste heat as a byproduct of its internal processes, the possibility also exists of harnessing this excess body heat and producing clean and usable energy. By utilizing a phenomenon known as the Seebeck Effect, a suitable apparatus may convert this excess thermal energy directly into electrical energy in order to produce a usable electrical current (Decher, 1997).

Literature Review

Physics of Heat

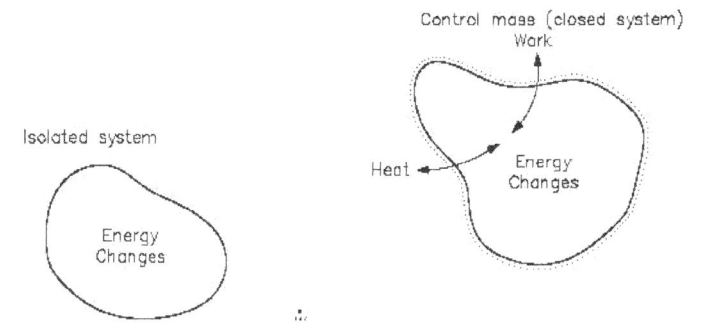

Figure 1. An illustration depicting two thermodynamic systems. In the isolated system, all transference of energy occurs within the system boundaries, and the amount of energy in the system remains constant. In the closed system, the mass of the system is held constant, but energy in the form of heat is introduced to the system and allowed to escape, causing the amount of energy contained within the system to change as work is done ("Thermodynamics, heat transfer, and fluid flow," 1992).

Heat is defined as the transfer of thermal energy from one medium or location to another in a system as a result of a difference in temperature between the two mediums. The Kinetic Molecular Theory of matter states that thermal energy results from the kinetic energy and consequent motion of the molecules within that object. When these molecules collide with those of another object, their kinetic and thus thermal energy is transferred to that object. This method of heat transfer by direct contact is known as conduction. Another primary method of heat transfer is radiation, or heat exchange without direct contact. One method of inducing thermodynamic heat transfer throughout a system is by means of a heat exchanger that crosses two fluid streams of different relative temperatures. The higher temperature fluid stream transfers heat energy to the lower temperature stream by conduction because of the principal or Zeroth Law of Thermodynamics (Kaminski & Jensen, 2005). A heat sink can also effect a dissipation of heat from a given thermodynamic system and prevent its overheating. Because the potential for radiative heat transfer increases with the surface area of a material and conductive heat transfer depends upon its thermal conductivity, such heat sinks consist of materials conducive to thermal energy designed in such a way as to maximize their surface area. Thus, commercial thermal heat sinks often employ a fin-like structure designed to dissipate unwanted heat with greater efficiency (Spakovszky, 2002).

The principles of heat are governed by the laws of thermodynamics, the overarching principal of which states that a system is always moving towards equilibrium, and heat will be transferred as long as there are differences in temperature within the system. In addition, the First Law of Thermodynamics states that energy is always conserved in an isolated system, in which no energy or mass enters or escapes. Thus, no energy in the form of heat can be destroyed in a system but only transferred or converted to other forms. This law can be seen demonstrated in Figure 1 above. Additionally, the second law states that work must be applied in a system in order to effect a change from a relatively cold to relatively heated state (Kaminski & Jensen, 2005).

Hence, because work is being applied within the system, this work can be utilized and converted to other forms of energy. Furthermore, because thermal energy results from the kinetic energy of the particles of a substance, any transfer of heat can conceivably be used to produce usable kinetic energy (Kaminski & Jensen, 2005). However, although the presence or lack of a temperature differential will determine whether thermal energy will be transferred in a system, temperature is not entirely indicative of thermal energy. Each material substance possesses a unique specific heat, which is a measure of the amount of energy required to raise a given volume of the material by one unit of temperature.

Physics of Electricity

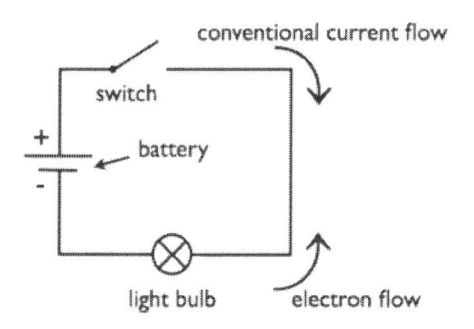

Figure 2. A diagram of a simple electrical circuit. The circuit comprises a voltage source, a conductor, a load, and a control. The flow of electrons in the circuit is shown to be opposite that of the conventional representation of electrical current flowing from positive to negative (Kightly, n.d.)

Electrical energy results from the transfer of electrical charge in a circuit. In general, electricity in a conductive circuit will flow from the location of highest potential voltage to that of lowest potential, and electrical energy is essentially the result of the flow of negatively charged electrons in this assembly.

The necessary components of an electrical circuit in order to allow for the transfer of electrical energy, as illustrated in Figure 2, are a source of voltage, such as a battery, and a conductive material, often copper or silver, connecting two locations of different electrical potentials and surrounded by an insulating material, often rubber. An electrically conductive material has a relatively low resistance to the flow of electricity with itself as a medium, whereas an electrical insulator is a material that is relatively resistant to the flow of electricity. Air, for example, often acts as an insulator, inhibiting or preventing the free flow of electricity (Gibilisco, 2001).

This flow of electricity through a circuit is referred to as an electrical current. In addition, an electrical circuit will often incorporate a load, which will increase the resistance to the flow of electricity through that circuit (Gibilisco, 2001). The current (I) in a completed electrical circuit is directly proportional to the voltage (V) and inversely proportional to the resistance (R) of the circuit, such that $V = IR$, with voltage measured in volts, current measured in amperes, and resistance measured in ohms. Furthermore, any circuit that possesses a measurable voltage and current also supplies electrical power (P). The electrical power in a circuit is directly proportional to both the voltage and current of that circuit, such that P = IV with power measured in watts (W), current measured in amperes, and voltage measured in volts.

Direct current (DC) is a form of electrical current that differs from alternating current (AC) in that it flows in only one direction in an electrical circuit. Many devices can be operated using DC current but require a certain voltage to operate effectively. However, utilizing a capacitor, a device that stores and periodically emits voltage, even a low-voltage DC current can be utilized to produce a more substantial and usable voltage and current that will apply power to standard DC electrical devices (Fleming, 2010).

DC-to-DC Conversion

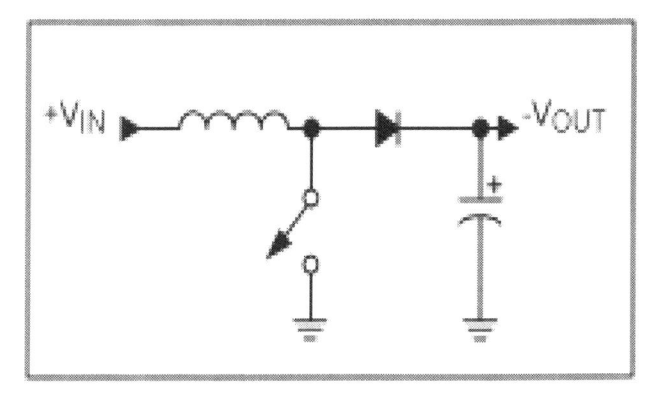

Figure 3. Circuit diagram of a simple DC-to-DC boost converter. The input voltage (V_{IN}) is stored using a capacitor during the charging phase and then discharged as a higher output voltage (V_{OUT}) ("Simple boost converter," 2001).

A DC-to-DC converter is any integrated circuit that accepts a DC voltage as input and outputs a DC voltage or current that differs from that of the input. Two common types of DC-to-DC converters are the buck, or step-down converter that reduces the voltage of the input and the boost, or step-up converter that effectively increases the input voltage. Circuits under this description operate in two distinct stages: the charging phase and the discharge phase. During the charge phase, a closed switch grounds the circuit and induces a voltage distribution that in turm charges a capacitor. During the discharge phase, this switch is opened and this capacitor discharges its stored voltage as an increased output voltage. Because power is conserved in a conversion circuit, the output current will be lower than the input provided to the circuit, and because of the inefficiency of circuit components, the final power of output of the circuit will be slightly lower than the input power ("DC-DC converter tutorial," 2001).

A charge pump, as seen in Figure 4 below, is a form of DC-to-DC boost converter that incorporates multiple capacitors and multiphased switches set to wire the capacitors in parallel during the charging phase and in series during the discharge phase in order to compound their individual stored voltages (Fleming, 2010).

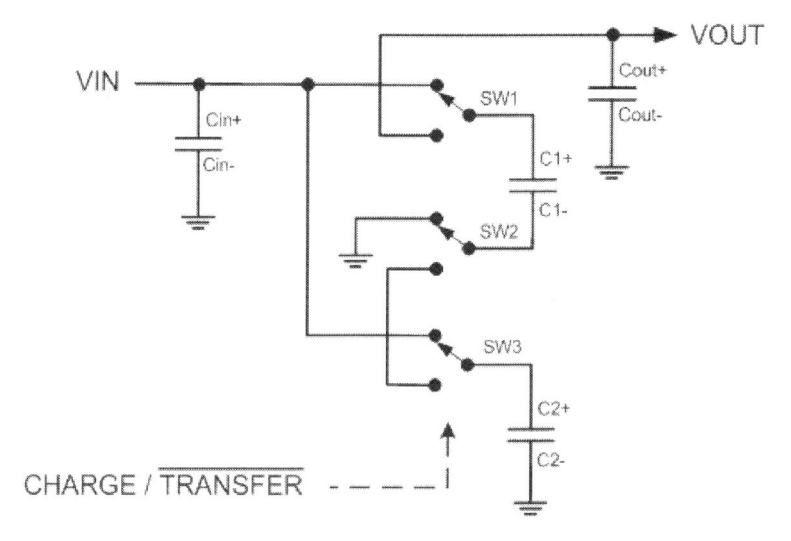

Figure 4. Circuit diagram of a simple DC-to-DC voltage doubling charge pump. The three switches (SW1, SW2, and SW3) alternate between their respective nodes in the circuit in order the provide voltage to and discharge it in series from the three circuit capacitors (Fleming 2010).

According to Prof. Alexander Emanuel, a professor of electrical engineering at the Worcester Polytechnic Institute in Massachusetts, these charge pumps are ideal for applications involving low-power source voltages because of their simple design and high efficiency (personal communication). They are also quite versatile, as a single charge pump can be combined with others in an array in order to produce a variety of voltage outputs (Fleming, 2010).

Potential of Unharnessed Waste Heat

Table 1. Measurements of the surface areas of, temperatures of, and heat lost by various parts of the human body (Cross, Collard & Nelson, 2008).

Segment	SA	DD	Tsk20	HL20	Tsk25	HL25	Tsk30	HL30	Tsk35	HL35
Upper arms	1996.13	169.53	28.0	24.62	30.8	17.85	33.4	10.46	36.1	3.39
Lower arms	1170.83	202.56	27.7	15.11	30.3	10.40	33.6	7.06	35.8	1.57
Hands	842.93	241.46	24.0	5.85	25.4	0.60	32.9	4.24	35.9	1.32
Upper legs	3077.00	167.34	27.9	37.10	30.5	25.84	33.4	15.96	35.1	0.47
Lower legs	2635.1	166.09	25.8	25.13	28.9	16.84	32.7	11.66	35.4	1.73
Feet	1358.63	164.13	21.7	3.52	27.1	4.34	34.8	9.93	35.6	1.24
Head and neck	1631.2	154.75	32.9	141.85	33.9	99.80	34.8	55.65	35.9	7.98
Trunk	6432.52	155.54	31.3	141.85	33	99.80	34.5	55.65	35.6	7.98

SA = surface area. DD = displacement distance during one cycle. Tsk20 = skin temperature when ambient temperature is 20°C. HL20 = heat loss at 20°C. Tsk25 = skin temperature when ambient temperature is 25°C. HL25 = heat loss at 25°C. Tsk30 = skin temperature when ambient temperature is 30°C. HL25 = heat loss at 30°C. Tsk35 = skin temperature when ambient temperature is 35°C. HL35 = heat loss at 35°C. Surface areas in cm^3. Displacement distances in cm. Skin temperatures in °C. Heat loss values in Watts.

In order to operate the muscles the human body converts stored chemical potential energy into mechanical energy (Shitzer & Eberhart, 1985a). Like any device that converts one form of energy to another, the muscles within the human body are not perfectly efficient, and some energy is lost in the form of heat. The muscular system of the body is able to convert chemical potential energy to mechanical energy with an average efficiency of 25%, with the remaining 75% of the energy converted to waste heat, which is then transferred by conduction to the circulatory system and then to the surface of the skin where it is transferred to the surrounding air ("Body heat to power cell Phones," 2008).

Hence, the excess heat produced by the muscles within the human body can be harnessed and converted as can waste heat given off by an inefficient combustion engine. The evidence in Table 1 indicates that of all the segments of the human body from which heat is lost to the ambient air, the human head, neck, and upper torso produce the majority of the heat that is lost (Cross, Collard & Nelson, 2008).

Other biological processes that cause a dissipation of excess heat from the human body include those of respiration and perspiration, both of which are essential to prevent the overheating of the body. Heat energy transferred away from the system of the body in these ways cannot be as easily harnessed and subsequently utilized. Furthermore, any condition that could cause an interruption of these processes would pose a serious and immediate risk of an individual's overheating and suffering from a heat-related medical condition such as heatstroke (Shitzer & Eberhart, 1985b).

Powering Electrical Devices

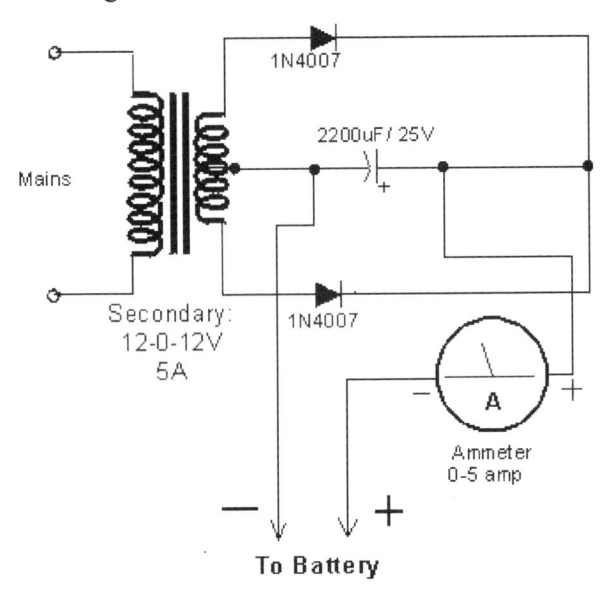

Figure 5. A battery charging circuit. The circuit takes a 12 V, 5 A electrical input current and produces an electrical current sufficient to provide a voltage charge to an external car battery, utilizing diodes and capacitors to regulate the flow of electricity ("Car battery charger circuit," 2010).

    A battery charging circuit, like the one shown above in Figure 4, works on the basic principle of reversing the flow of electrical energy in a circuit to which a battery is connected by providing an outside voltage source. If the voltage provided is sufficient, the potential energy of the battery can be restored (Brain & Bryant, 2000).

    Rechargeable batteries often consist of electrically conductive materials such as nickel, cadmium, and lithium, but vary in design and requirements in terms of voltage and current. An electrical power converter can be incorporated into a battery charging circuit in order to provide sufficient electrical energy to suit the unique requirements of a given battery. The duration required to fully recharge a battery with modern recharging devices also varies, falling into one of three categories: slow charging speed, which requires 14 to 16 hours; average charging speed, which requires 3 to 7 hours; and fast charging speed, which requires no more than one hour (Brain & Bryant, 2000).

A 3.7 V lithium-ion battery provides the electrical current that powers most standard video iPod devices. The battery is designed to be rechargable by application of a 5 or 12 V current. A standard iPod charger will draw power from most standard household electrical outlets and provide power sufficient to charge the battery (Layton, 2006). However, any device capable of producing a similar voltage and current could conceiveably provide the charge necessary to replenish the battery life of the iPod.

Hearing aid devices are typically powered by D batteries with a capacity of 1.5 V. Many household flashlights incorporate this same voltage battery. However, because hearing aids draw a large amount of power and are typically in constant use by hearing impaired individuals, their effective battery life is typically of short duration (Carmen, 2004).

Thermoelectric Circuit Design and the Thermoelectric

Effects

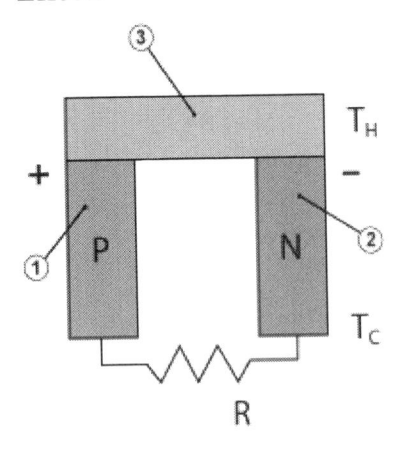

Figure 6. A visual diagram of a thermoelectric circuit. The materials labeled N and P represent two semiconductive component materials of a thermocouple to which voltage can be applied to induce a temperature differential at junction 3 or a heat source can be applied at junction 3 to cause a flow of electrons and thus induce an

As illustrated in Figure 3, a thermoelectric circuit comprises a thermocouple made up of two conductive or semiconductive n-type and p-type materials. The former materials possess an excess number of electrons and the latter possess an excess number of protons. These two materials are connected by a thermally conductive junction or junctions and an electrically conductive circuit, often with a load, such as a resistor (Decher, 1997).

A thermoelectric circuit can undergo either of two thermoelectric processes when provided with either electrical or heat energy. These processes are known as the thermoelectric effects. The first of these, the Peltier Cooling Effect, is a process by which electrical energy can be utilized to produce a temperature differential at the junction of the thermoelectric circuit and thus induce cooling. If an electrical voltage is supplied as an input to the circuit, the resulting flow of electrons through the circuit and thermocouple component materials will induce a temperature differential at either junction connecting the materials, cooling one of the junctions while heating the other (U.S. Patent No. 5,288,336, 1994). The reverse of this phenomenon is termed the Seebeck Effect and is a thermoelectric effect that can be utilized to convert heat energy into electrical output voltage. This occurs when a heat source is introduced to the junction of the thermocouple within the thermoelectric circuit. The transfer of heat through the semiconductive n- and p-type materials causes a flow of excess electrons from the n-type material. This flow of electrons through the circuit induces a current, the power of which is positively correlated with the amount of thermal energy that is introduced to the circuit (U.S. Patent No. 5,288,336, 1994).

Recent Work in the Field of Thermal to Electric Energy Conversion

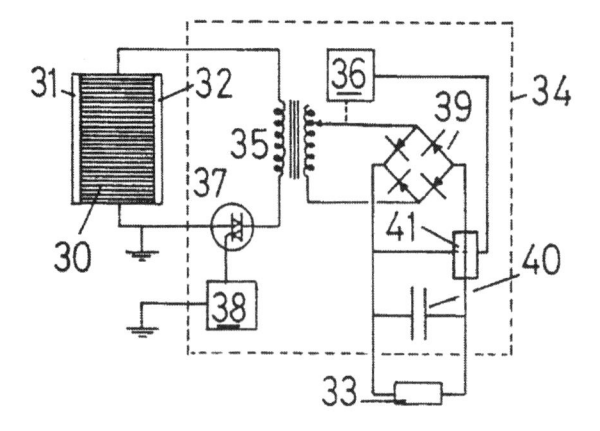

Figure 7. A 1994 patented design for a thermoelectric energy converter. The design utilizes the previously described thermoelectric circuit design and incorporates aluminum and nickel as materials in a thermocouple as well as a junction design meant to recapture heat lost from the thermocouple (U.S. Patent No. 5,288,336, 1994).

Until recently, many thermocouple components used in thermoelectric circuit designs comprised fully metallic conductors, but semiconductive n-type and p-type materials are now frequently used instead in order to produce energy with higher efficiency (Decher, 1997). Many modern researchers in the field of thermoelectric energy conversion have focused on isolating material components with a thermoelectric figure of merit (ZT) greater than 1.0. This unitless value directly corresponds to the efficiency of a thermoelectric device composed of this material and is calculated as the product of the square of the Seebeck coefficient of the material and the electrical conductivity of the circuit divided by its thermal conductivity.

Currently, no thermoelectric device exists with an energy conversion efficiency of greater than five percent when provided with a relatively low input energy in the form of heat resulting from a temperature differential of less than 200 K despite many attempts in the field (Rowe, 1999). For example, researchers tested a phenomenon known as thermotunneling as related to thermal transfer at the junctions of a thermoelectric generator in an attempt to improve the value of ZT. However, through experimentation the investigators found that the process produced an inherent limiting value for ZT that did not exceed the existing values achieved for such devices (Dillner, 2010). Instead, much research has been directed toward application of such a device to provide electrical output when provided with a temperature differential of 500 K or more. Such conditions yield energy conversion of higher efficiency, approximately 11% more efficient than at lower temperatures (Yamashita, 2009).

Previous research into the potential of thermoelectric conversion as a means of harnessing low amounts of waste heat to power electrical devices has yielded largely theoretical applications. For example, researchers investigated the potential of a thermoelectric device that would utilize the heat produced by the wrist of a wearer to power a small wristwatch. However, the output voltage of such a device was calculated to be very low—in figures of millivolts—on any reasonable scale because of the low efficiency of the device, and as such, the design failed to yield any practical applications (Rowe, 1999).

The U.S. Patent office has issued multiple patents to devices that claim to improve upon the design and efficiency of existing thermoelectric energy conversion devices by using various materials to construct the thermocouple component. One of these designs, incorporating nickel and aluminum as thermocouple materials, can be seen in Figure 5. In addition, experts in the field of thermoelectric energy have conducted much research into the topic in recent years and propose that materials such as silicon fibers may be incorporated into the thermocouple of such a device in order to further improve its figure of merit to a value greater than 1.0. The conversion efficiency of such an apparatus may prove sufficient to render the utilization of waste body heat to power electrical devices such as cellular phones plausible. However, such technology and materials remain expensive and thus not readily available or marketable on any practical scale ("Body heat to power cell phones," 2008). In a 2010 study, researchers fabricated a series of small-scale thermoelectric generators and adhered them to the body of various test subjects. The investigators composed the the generators using silicon-germanium and bismuth telluride as n- and p- materials. By wiring 1300 of these generators in series, the researchers measured output voltages between 0.15 and 0.70 V using only the difference between the subjects' body temperature and the ambient temperature of 22°C (Su et al.,

2010).

Research Proposal

Problem

A problem with the inefficiency of current methods of energy conversion has led to difficulty in harnessing the heat lost by the human body.

Engineering Goal

The goal of this project is to design a prototype marketable device that will be worn on the body and will convert lost heat into electrical energy sufficient to power a small electrical circuit or device, such as a standard video iPod.

Methods and Procedure

This device will be designed and incorporated into an article of clothing, such as an insulated cap, and will convert excess body heat into electrical energy. The device will comprise a thermoelectric circuit and utilize the Seebeck Effect to harness the heat resulting from the temperature differential between the epidermis and an ice water-cooled junction to produce a low-power electrical current. As predicted by electrical engineer Carl Guyer, the voltage produced by this arrangement is likely to be no more than several to tens of millivolts (personal communication). Therefore, an ultra-low DC-to-DC voltage converter will then harness this energy to produce a voltage of approximately 5 to 12 volts (V), sufficient to power a standard iPod.

In a procedure recommended by Carl Guyer, the apparatus

will first be tested using a controlled, non-human source of heat, such as a hot water bottle, and the electrical output of the system will be measured in order to test its practicality and efficiency (personal communication). It will then be incorporated into an insulated article of clothing, worn and tested by the author in different temperature environments, during periods of idle rest and heightened activity, such as jogging. The values obtained for the electrical output will then be compared against power requirements for common electrical devices, such as an iPod and hearing aid, and also against the power consumption of these devices. Additional engineering goals include powering a hearing aid or other devices that require more power to operate than a standard iPod model and designing the apparatus to be optimally cost-effective and efficient. The device will also be made as plausibly marketable as possible by minimizing its defects and optimizing its practicality as a personal electrical energy generator.

Methodology

Experiment 1: Temperature of Cranial and Thoracic Epidermis in Various Conditions

The ambient air temperature was first measured using a standard mercury thermometer. Throughout the experiment, the investigator was clothed in a standard 100% cotton short-sleeve shirt worn underneath a standard 100% cotton long-sleeve shirt and jeans (Levi's, W 28 L 32, slim straight men's, model 514). A primary hat (Blank, long cuff winter beanie, 96% acrylic) was placed onto the head of the investigator, covering the eyebrows and one centimeter below the lobes of the ears. Then, with the investigator at idle rest the epidermal surface temperature was measured directly above the right ear tip, directly below the right ear lobe, one centimeter above the rear hemline of the hat, and at the rear base of the neck using the thermocouple probe attachment of an infrared temperature sensor (Craftsman, -29 to 760°C, No. 0808869). These epidermal surface temperature measurements were taken periodically at five-minute intervals for a duration of one hour using the thermocouple probe. At each five-minute interval, the ambient air temperature was also measured using the standard mercury thermometer. After this, the hat was removed by the investigator, who then remained at rest in the ambient temperature environment for an additional hour. The above experimental procedure was then repeated with the primary hat replaced by a secondary hat (Eastern Mountain Sports, Atlas Fleece Hat, EMS #3938790) and again without a hat in order to establish a control for epidermal surface temperature without insulation. Then the entirety of the above experimental procedure was repeated with the investigator actively jogging on a treadmill (Proform, model 840, PFTL721040) at a speed of approximately 6.44 kph and an incline setting of one percent while epidermal surface temperature measurements were taken.

Experiment 2: Thermoelectric Generator Performance with Controlled Heat Source

First, the surface temperature of the external junction of a thermoelectric generator assembly (Hi-Z, bismuth telluride, water cooled, model HZ-20) was measured with the infrared temperature sensor. A hot water bottle (Rite Aid, Deluxe Hot Water Bottle, rubber, 1.89 L capacity) was then filled with 1.5 L of water heated to 34.0°C. The air was expelled from the bottle by the application of a slight compression force to the dorsal and ventral sides of the bag until a small pool of water formed in the neck of the bottle. The stopper was then inserted into the bottle and twisted clockwise to seal the bag before the compression force was removed. The hot water bottle was placed on a flat maple wood surface, and its central dorsal surface temperature was measured using the infrared temperature sensor. Then the ambient air temperature was measured with the standard mercury thermometer, and the thermoelectric generator was placed onto the center of the water bottle, such that its external junction was flush with the dorsal surface of the bottle. The peak voltage output of the generator and the peak current drawn from the electrical circuit with a 5 Ω external resistance were measured with a standard electric multimeter.

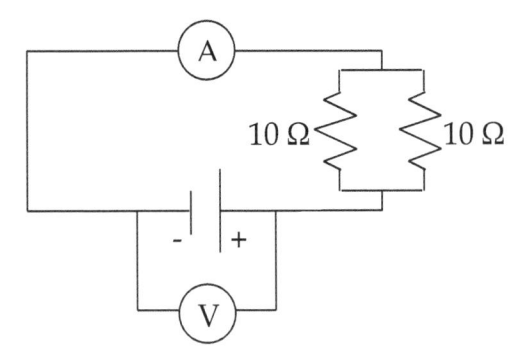

Figure 7. A circuit diagram of the setup used to measure the voltage and current output of the thermoelectric generator in Experiment 2. Voltage is measured parallel to the positive and negative output terminals, and a five Ω resistance is achieved by wiring two standard 10 Ω resistors in parallel in order to draw the maximum possible current via load matching.

The generator was then removed from the bottle and the surface temperature of the external junction allowed to return to its baseline measured value. The above experimental procedure was repeated for four additional trials with hot water heated to 34°C and for five trials with water heated to temperatures of 37°C and 40°C as well as three trials with the generator placed on an insulated cotton surface at ambient air temperature in order to establish a control for measuring the electrical output of the generator.

The entirety of the above procedure was then repeated twice more with a medical enema bag and tube apparatus (Medline, enema kit, DYND70100) attached to a plastic water bottle (Nalgene, one liter capacity, HS340747) supplying the generator with a constant flow of distilled water cooled to a temperature between 0°C and 2°C. Five trials were performed for each hot water temperature setting, both without restriction of the flow of ice water and with a partial restriction placed on this flow. This restriction was achieved by setting the plastic crimping tab attached to the tube of the medical enema bag to its central lengthwise position on the tube. The temperature of the cooled water was measured prior to each trial with a digital cooking thermometer (Taylor, TruTemp 3519, -40°C to 230°C). The water flow was powered by liquid suction induced by manually siphoning the water through the tube and concurrently halting the flow of water by crimping the tube completely with the attached plastic crimping tab. The height difference between the plastic bottle and bag allowing for liquid suction was measured to be approximately 46 cm with a six cm margin of error as a result of the change in water level over the course of the experiment.

Prototype A Assembly – Acrylic Hat with Thermoelectric Generator Apparatus

A uniform six centimeter by six centimeter square hole was cut in the center of the apical surface of the acrylic hat from Experiment 1. The thermoelectric generator was aligned with this hole and placed flush with the apical surface of the hat such that the external junction of the generator protruded through the hole and to the interior of the hat. Four *25⁄16* cm-diameter holes were bored through the corner extremities of the dorsal and ventral surfaces of the generator *5⁄16* cm from each of the nearest edges of the surface. The generator was then affixed to the hat by sewing the four holes to the permeable apical surface of the hat with thread (Talon, 30/2 weight, 100% cotton).

Prototype B Assembly – Fleece Hat with Thermoelectric Generator and Cooling System

A uniform six centimeter by six centimeter square hole was cut in the center of the apical surface of a hardhat liner (Ergodyne, two-layer, 100% cotton twill shell, 100% poly-fleece lining, 16850EG). The thermoelectric generator was aligned with this hole and placed flush with the apical surface of the liner such that the external junction of the generator protruded through the hole and to the interior of the hat. The generator was then affixed to the liner by sewing the four pre-bored holes from the assembly of Prototype A to the permeable apical surface of the liner with the cotton thread used in Prototype A. Two leads of standard rubber-insulated copper wires were affixed to the negative output terminal of the generator using the attached screws. An additional lead was affixed to the positive output terminal of the generator and a third was soldered to a parallel circuit comprising two standard 10 Ω resistors in parallel which was in turn attached to the positive terminal of the generator. The fleece hat from Experiment 1 was then placed so that its interior surface enveloped the liner on which the generator was mounted and its hemline came to one centimeter below the tip of both ears of the investigator. Two of the plastic one liter bottles from Experiment 2 were placed in the pouches of a backpack (SwissGear, Granite, GA-7335-07F00) such that the height from the top of the upper bottle to the top of the uppermost handle of the backpack was nine cm and the height difference between the top of the upper bottle and the top of the lower bottle was 20.5 cm. Two lengths of the tube apparatus used in Experiment 2, measuring 90 cm and 80 cm, were attached to the liquid input and output nozzles of the generator. The open end of the 90 cm length of tube was placed into the lower bottle on the backpack, and the 80 cm length of tube was placed into the upper bottle.

Prototype C Assembly – Hardhat Liner with Three Generators and Cooling System

Additional six centimeter by six centimeter square holes were cut in the sides of of the hardhat liner from the Prototype B assembly. Two additional thermoelectric generators were aligned with these holes and placed flush with the surface of the liner such that the external junction of the generator protruded through the hole and to the interior of the hat. To each generator was then affixed a water cooling block (Enzotech, copper, SCW-REV.A) with thermal adhesive, and each was then attached to the liner by wrapping thread around their centers and sewing them to the permeable surface of the liner with the cotton thread used in Prototypes A and B. Leads of standard rubber-insulated copper wires were soldered to those of the other generators in order to connect them in a series circuit. An additional lead was affixed to the negative output terminal of the generator and connected to a 0.6 Ω equivalent resistor in an external circuit, which matched the impedance of the generators. Four lengths of the tube apparatus used in Experiment 2 were used to attach to the liquid input and output nozzles of the generators in a series. The open ends of tubes were placed into the bottles used with Prototype B.

Experiment 3: Prototype Performance Testing
The surface temperature of the external junction of the thermoelectric generator of Prototype A was first measured using the infrared temperature sensor. Prototype A was then placed onto the head of the investigator, and the ambient air temperature was measured with the standard mercury thermometer. The standard multimeter was used to measure the voltage and current output of the device as outlined in the procedure for Experiment 2. The prototype was then removed from the bottle and the surface temperature of the external junction allowed to return to its baseline measured value. Then the prototype was placed on an insulated cotton surface at ambient air temperature and its voltage and current output were again measured with the standard multimeter in order to establish a control for the electrical output of the generator without exposure to heating as a result of metabolic heat generation. The above procedure was repeated for seven additional trials.

The entirety of the above procedure was then repeated, substituting Prototypes B and C for Prototype A and allowing ice water to flow through the generator for each trial using the procedure outlined in the procedure for Experiment 2, both with and without the prototype resting on the head of the investigator. Before each trial, the temperature of the ice water was measured using the digital cooking thermometer.

Results

Experiment 1:

Table 2. Cranial and thoracic epidermal surface temperature measurements with varied subject activity and hat materials.

Activity	Material	Time (min)	Ambient Temp. (°C)	Body Area	Surface Temp. (°C)
None	Control	0	19.8	Front of Ear	31.5
				Top of Head	30.1
				Front of Temple	32.3
				Neck/Torso	31.5
		5	20.3	Front of Ear	32.2
				Top of Head	30.9
				Front of Temple	32.4
				Neck/Torso	31.7

		10	19.5	Front of Ear	32.0
				Top of Head	31.0
				Front of Temple	32.2
				Neck/Torso	31.5
		15	20.1	Front of Ear	32.1
				Top of Head	31.3
				Front of Temple	32.5
				Neck/Torso	31.6
		20	19.9	Front of Ear	32.2
				Top of Head	31.4
				Front of Temple	31.9
				Neck/Torso	31.4
		25	20.0	Front of Ear	32.2
				Top of Head	31.6
				Front of	32.0

				Temple	
				Neck/Torso	31.4
		30	19.2	Front of Ear	32.1
				Top of Head	31.6
				Front of Temple	31.7
				Neck/Torso	31.4
		35	19.7	Front of Ear	32.0
				Top of Head	31.6
				Front of Temple	32.0
				Neck/Torso	31.8
		40	20.3	Front of Ear	31.9
				Top of Head	31.4
				Front of Temple	31.7
				Neck/Torso	31.3
		45	19.8	Front of Ear	32.0
				Top of	31.6

				Head	
				Front of Temple	31.7
				Neck/Torso	31.3
	Acrylic	0	20.3	Front of Ear	32.1
				Top of Head	31.0
				Front of Temple	32.8
				Neck/Torso	33.2
		5	19.7	Front of Ear	33.0
				Top of Head	31.2
				Front of Temple	33.4
				Neck/Torso	33.5
		10	19.0	Front of Ear	33.5
				Top of Head	30.7
				Front of Temple	34.0
				Neck/Torso	33.5
		15	20.5	Front of Ear	33.5

				Top of Head	31.0
				Front of Temple	33.9
				Neck/Torso	33.3
		20	19.2	Front of Ear	34.0
				Top of Head	31.1
				Front of Temple	33.9
				Neck/Torso	33.2
		25	19.9	Front of Ear	34.2
				Top of Head	32.0
				Front of Temple	34.3
				Neck/Torso	33.3
		30	19.8	Front of Ear	34.0
				Top of Head	32.2
				Front of Temple	34.3
				Neck/Torso	33.4
		35	19.3	Front of	34.0

				Ear	
				Top of Head	31.8
				Front of Temple	34.5
				Neck/Torso	33.7
		40	19.7	Front of Ear	34.0
				Top of Head	31.9
				Front of Temple	34.3
				Neck/Torso	33.8
		45	19.2	Front of Ear	34.2
				Top of Head	31.7
				Front of Temple	34.5
				Neck/Torso	33.9
	Fleece	0	20.2	Front of Ear	32.4
				Top of Head	31.4
				Front of Temple	32.8
				Neck/Torso	33.2

		5	19.0	Front of Ear	32.7
				Top of Head	31.2
				Front of Temple	33.6
				Neck/Torso	33.1
		10	19.9	Front of Ear	33.6
				Top of Head	31.1
				Front of Temple	34.1
				Neck/Torso	33.4
		15	19.5	Front of Ear	33.7
				Top of Head	31.9
				Front of Temple	34.0
				Neck/Torso	34.1
		20	19.9	Front of Ear	34.0
				Top of Head	32.2
				Front of Temple	34.4

				Neck/Torso	33.7
		25	19.1	Front of Ear	34.2
				Top of Head	32.3
				Front of Temple	34.4
				Neck/Torso	34.3
		30	19.8	Front of Ear	34.2
				Top of Head	32.4
				Front of Temple	34.5
				Neck/Torso	34.0
		35	20.0	Front of Ear	34.7
				Top of Head	32.6
				Front of Temple	34.3
				Neck/Torso	34.0
		40	20.3	Front of Ear	34.2
				Top of Head	32.3
				Front of	34.7

				Temple	
				Neck/Torso	34.3
		45	19.2	Front of Ear	34.3
				Top of Head	32.9
				Front of Temple	34.6
				Neck/Torso	34.6
Jogging	Control	0	20.1	Front of Ear	31.1
				Top of Head	29.5
				Front of Temple	33.1
				Neck/Torso	33.9
		5	19.7	Front of Ear	32.3
				Top of Head	30.0
				Front of Temple	32.4
				Neck/Torso	33.0
		10	19.3	Front of Ear	33.2
				Top of Head	31.1

				Front of Temple	33.5
				Neck/Torso	33.3
		15	19.8	Front of Ear	33.7
				Top of Head	31.2
				Front of Temple	33.0
				Neck/Torso	33.1
		20	20.2	Front of Ear	33.6
				Top of Head	31.9
				Front of Temple	32.9
				Neck/Torso	33.2
		25	19.6	Front of Ear	33.8
				Top of Head	32.1
				Front of Temple	33.2
				Neck/Torso	33.3
		30	19.5	Front of Ear	34.1

		Top of Head	32.2
		Front of Temple	33.4
		Neck/Torso	33.4
35	19.8	Front of Ear	33.9
		Top of Head	32.1
		Front of Temple	33.5
		Neck/Torso	33.3
40	20.0	Front of Ear	34.2
		Top of Head	32.7
		Front of Temple	33.7
		Neck/Torso	33.5
45	20.4	Front of Ear	34.0
		Top of Head	32.9
		Front of Temple	33.7
		Neck/Torso	33.4

	Acrylic	0	21.2	Front of Ear	32.2
				Top of Head	31.0
				Front of Temple	33.6
				Neck/Torso	32.1
		5	21.0	Front of Ear	32.8
				Top of Head	32.3
				Front of Temple	33.5
				Neck/Torso	32.4
		10	20.7	Front of Ear	33.5
				Top of Head	31.8
				Front of Temple	33.7
				Neck/Torso	32.5
		15	19.8	Front of Ear	32.6
				Top of Head	31.9
				Front of Temple	33.9

			Neck/Torso	32.7
	20	20.2	Front of Ear	33.6
			Top of Head	31.7
			Front of Temple	33.5
			Neck/Torso	32.8
	25	19.7	Front of Ear	33.7
			Top of Head	31.9
			Front of Temple	34.2
			Neck/Torso	32.5
	30	21.0	Front of Ear	34.1
			Top of Head	32.2
			Front of Temple	34.3
			Neck/Torso	32.6
	35	20.0	Front of Ear	33.9
			Top of Head	32.0
			Front of	34.5

				Temple	
				Neck/Torso	32.8
		40	20.5	Front of Ear	34.3
				Top of Head	32.4
				Front of Temple	34.6
				Neck/Torso	32.7
		45	20.3	Front of Ear	34.5
				Top of Head	32.9
				Front of Temple	34.9
				Neck/Torso	33.1
	Fleece	0	20.6	Front of Ear	32.5
				Top of Head	31.7
				Front of Temple	33.2
				Neck/Torso	32.3
		5	20.0	Front of Ear	33.1
				Top of Head	31.9

				Front of Temple	33.9
				Neck/Torso	32.8
		10	19.3	Front of Ear	33.7
				Top of Head	32.1
				Front of Temple	34.6
				Neck/Torso	32.5
		15	19.7	Front of Ear	34.0
				Top of Head	32.9
				Front of Temple	34.7
				Neck/Torso	32.3
		20	20.0	Front of Ear	34.5
				Top of Head	33.5
				Front of Temple	34.8
				Neck/Torso	33.5
		25	20.6	Front of Ear	34.4
				Top of	33.3

			Head	
			Front of Temple	34.6
			Neck/Torso	33.1
	30	21.0	Front of Ear	34.5
			Top of Head	33.0
			Front of Temple	34.9
			Neck/Torso	32.9
	35	20.4	Front of Ear	34.7
			Top of Head	32.8
			Front of Temple	34.8
			Neck/Torso	33.3
	40	19.8	Front of Ear	34.8
			Top of Head	32.9
			Front of Temple	35.0
			Neck/Torso	33.5
	45	20.2	Front of Ear	34.7

				Top of Head	32.7
				Front of Temple	35.2
				Neck/Torso	33.7

Experiment 2:

Table 3. Voltage and current output of generator assembly with controlled heat source and cooling conditions. Temperature difference is the calculated absolute difference between the surface & ice water temperatures.

Cooling System	Hot Water Temp (°C)	Ambient Temp (°C)	Surface Temp (°C)	Ice Water Temp (°C)	Temp Difference (°C)	Voltage (mV)	Current (µA)	Power (mW)
Full Flow	34	19.2	31.1	1.7	29.4	43.5	13	0.566
		19.2	31.3	1.7	29.6	45.0	13.6	0.612
		18.8	31.5	1.0	30.5	50.7	15.3	0.776
		18.6	32.3	0.8	31.5	48.4	14.3	0.692
		20.3	30.5	1.6	28.9	45.6	13.6	0.620
	37	19.3	33.6	0.9	32.7	50.8	15.1	0.767
		18.7	33.5	1.9	31.6	49.6	15	0.744
		19.4	33.8	0.4	33.4	51.9	15.8	0.820
		18.8	33.3	1.0	32.3	52.3	16.2	0.847
		18.5	33	1.7	31.3	49.3	14.5	0.71

								5
	40	19.6	34.7	1.6	33.1	52.8	16.5	0.871
		18.8	35.4	1.2	34.2	56.4	17.8	1.004
		19.0	35	1.0	34.0	54.1	17.1	0.925
		18.6	35	1.7	33.3	57.1	17.5	0.999
		18.9	34.5	1.2	33.3	52.0	15.6	0.811
	Control	18.5	N/A	1.3	17.2	23.3	4.2	0.098
		19.0	N/A	0.8	18.2	25.1	6.1	0.153

		18.9	N/A	1.1	17.8	23.9	4.8	0.115
		19.1	N/A	0.9	18.2	21.9	3.7	0.081
		19.0	N/A	1.4	17.6	24.2	5.1	0.123
Crimped Flow	34	18.8	29.5	1.4	28.1	44.3	13.3	0.589
		19.2	31	0.9	30.1	47.9	14.2	0.680
		19.0	31.3	1.0	30.3	48.0	14.2	0.682
		18.5	29	1.3	27.7	40.0	10.5	0.420
		18.7	31	1.5	29.5	42.3	12.1	0.512
	37	19.4	33	1.0	32.0	49.4	14.4	0.711
		19.8	33.3	1.5	31.8	49.5	14.3	0.708
		19.2	33.7	0.8	32.9	50.6	14.8	0.749
		18.6	32.7	0.8	31.9	47.0	13.9	0.653
		18.4	33.8	1.9	31.9	47.3	14.1	0.667
	40	18.8	35.1	1.7	33.4	51.7	15.7	0.812
		19.0	34.8	0.9	33.9	53.1	16.6	0.881
		18.7	35.4	1.2	34.2	54.0	17.1	0.923
		18.3	35.5	0.3	35.2	54.1	17.3	0.936
		18.9	36.0	1.9	34.1	52.0	15.4	0.801
	Control	19.2	N/A	0.9	18.3	22.9	3.8	0.087
		18.5	N/A	0.7	17.8	22.4	3.6	0.081
		18.8	N/A	1.3	17.5	20.9	3.2	0.067
		18.4	N/A	1.0	17.4	20.7	3.3	0.068
		19.0	N/A	1.4	17.6	23.3	4.1	0.096
None	34	18.6	31.3	N/A	12.7	24.3	4.7	0.114
		19	30.5	N/A	11.5	23.5	3.7	0.087
		18.3	32.1	N/A	13.8	25.1	5.1	0.128
	37	18.4	33.8	N/A	15.4	25.8	5.4	0.139
		18.7	33.3	N/A	14.6	26.7	6	0.160
		19.3	32.8	N/A	13.5	25.4	5.3	0.135
	40	18.8	34.6	N/A	15.8	26.4	5.7	0.150
		19.4	35.5	N/A	16.1	27.3	6.2	0.169
		18.5	35	N/A	16.5	27	6.1	0.165
	Control	18.7	N/A	N/A	0.0	0.7	0.1	0.000
		19	N/A	N/A	0.0	0.3	0	0.000
		19.2	N/A	N/A	0.0	0.9	0.1	0.000

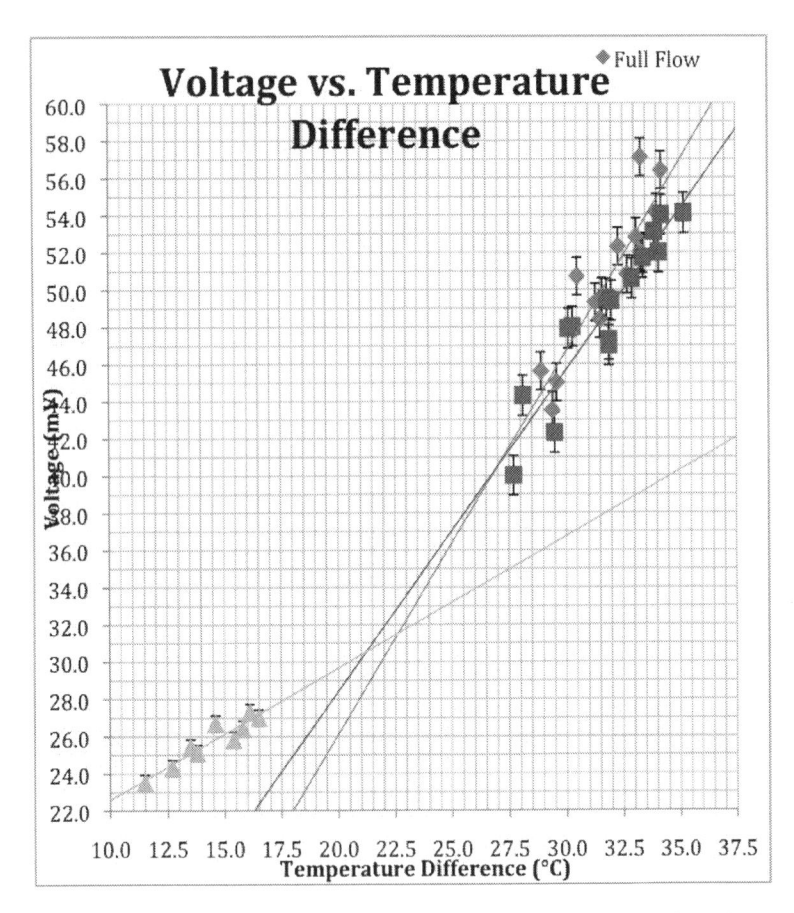

Figure 14. A graph of the voltage output of the thermoelectric generator as a function of the temperature difference between the surface temperature of the controlled heat source and the ice water cooling supply (or ambient temperature where applicable) for each of the three flow settings. The trend of voltage outputs of the unrestricted flow system was greater than that of the restricted flow system, both of which were substantially greater than the trend of the non-cooled system. All trends are modeled as linear, and standard deviation error bars are shown. Control results are not represented on this graph.

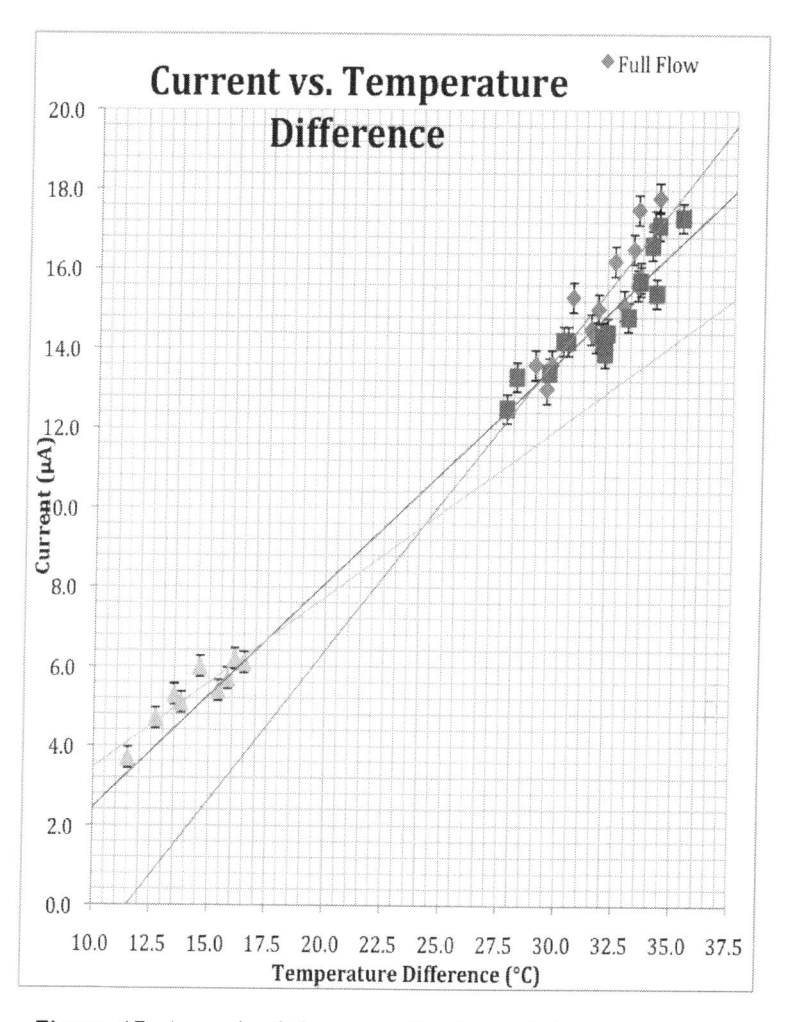

Figure 15. A graph of the current output of the thermoelectric generator as a function of the temperature difference between the surface temperature of the controlled heat source and the ice water cooling supply (or ambient temperature where applicable) for each of the three flow settings. The trend of current outputs of the unrestricted flow system was greater than that of the restricted flow system, both of which were greater than the trend of the non-cooled system. All trends are modeled as linear, and standard deviation error bars are shown. Control results are not represented on this graph.

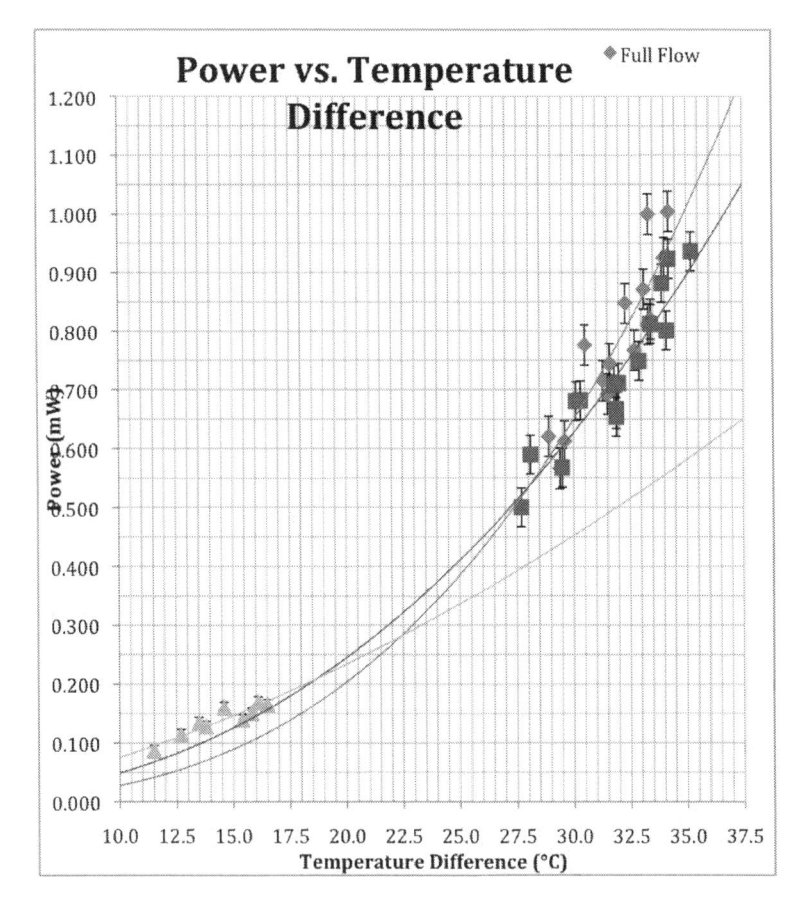

Figure 16. A graph of the power output of the thermoelectric generator as a function of the temperature difference between the surface temperature of the controlled heat source and the ice water cooling supply (or ambient temperature where applicable) for each of the three flow settings. The trend of current outputs of the unrestricted flow system was greater than that of the restricted flow system, both of which were greater than that of the non-cooled system. All trends are modeled as power functions, and standard deviation error bars are shown. Control results are not represented on this graph.

Experiment 3:

Table 4. Voltage and current output of final prototype

Prototype	Room Temp (°C)	Ice Water Temp (°C)	Voltage (mV)	Current (mA)	Power (mW)
A	18.9	0.9	33.6	7.0	0.235
	19.4	1.1	32.0	6.4	0.205
	19.0	1.4	31.2	6.3	0.197
	18.8	0.7	34.2	7.3	0.250
	19.2	0.5	33.9	7.2	0.244
	18.9	1.0	35.0	7.5	0.263
	19.0	1.4	30.8	6.0	0.185
	19.3	0.8	34.6	7.3	0.253
B	18.9	1.2	36.2	8.0	0.290
	18.6	1.0	36.5	8.2	0.299
	19.1	1.2	32.5	6.7	0.218
	19.4	0.6	37.0	8.5	0.315
	18.7	0.9	33.3	7.1	0.236
	18.8	1.1	34.6	7.4	0.256
	19.0	0.4	36.0	7.7	0.277

assemblies A, B, and C, and control (see Methodology).

| | 19.1 | 0.7 | 35.8 | 7.8 | 0.279 |

		18.6	1.2	82.5	9.9	0.817
		18.8	1.0	77.4	9.6	0.743
		19.1	1.2	80.3	10.3	0.827
C		19.0	0.6	79.8	9.7	0.774
		18.7	0.9	81.7	10.6	0.866
		19.3	1.1	82.6	10.8	0.892
		19.0	0.4	78.8	9.9	0.780
		19.4	0.7	80.8	10.3	0.832
		19.0	1.0	11.9	2.0	0.024
		19.6	0.8	24.7	5.6	0.138
		19.2	0.6	15.2	3.4	0.052
Control		18.7	1.1	14.2	3.2	0.045
		18.9	0.2	13.9	2.9	0.040
		19.0	1.2	13.4	2.5	0.034
		18.8	0.8	16.1	3.8	0.061
		19.0	1.0	12.7	2.3	0.029

Data Analysis and Discussion

Activity	Body Area	T-Test (Acrylic vs. Fleece)
None	Front of Ear	0.0957
	Top of Head	0.0015
	Front of Temple	0.0384
	Back of Neck/Torso	0.0203
Jogging	Front of Ear	0.0011
	Top of Head	0.0114
	Front of Temple	0.0066
	Back of Neck/Torso	0.0963
All		5.38×10^{-13}

Table 4. Results of Student T-tests performed to compare the measured average body temperatures during Experiment 1 for the acrylic and fleece hats.

Activity	Body Area	T-Test (Fleece vs. Control)
None	Front Ear	1.46×10^{-5}
	Top Head	2.29×10^{-4}
	Temple	1.94×10^{-5}
	Neck/Torso	5.39×10^{-7}
Jogging	Front Ear	5.30×10^{-5}
	Top Head	6.56×10^{-3}
	Temple	1.65×10^{-5}
	Neck/Torso	9.31×10^{-2}
All		3.70×10^{-17}

Table 5. Results of Student T-tests performed to compare the measured average body temperatures during Experiment 1 for the fleece hat and control.

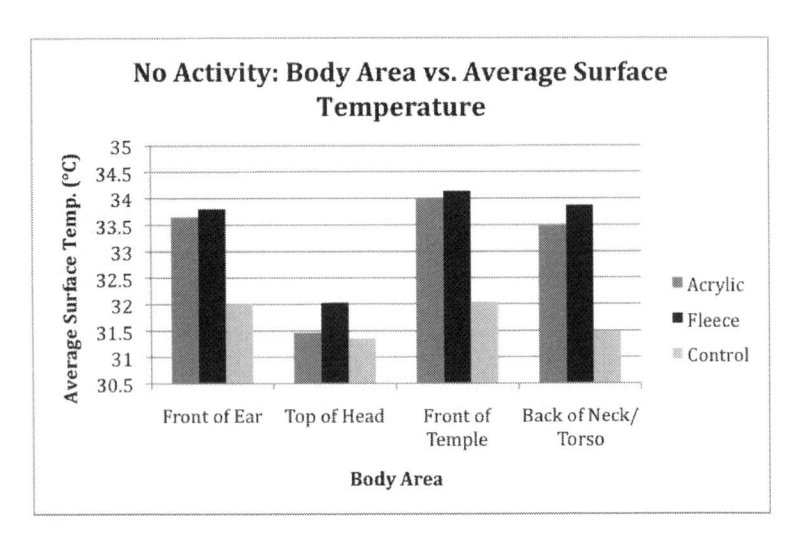

Figure 17. A graphical comparison, between acrylic and fleece hats and the control, of the average body temperatures achieved at various parts of the body while at rest. The fleece hat is shown to result in highest average body temperatures for all measured

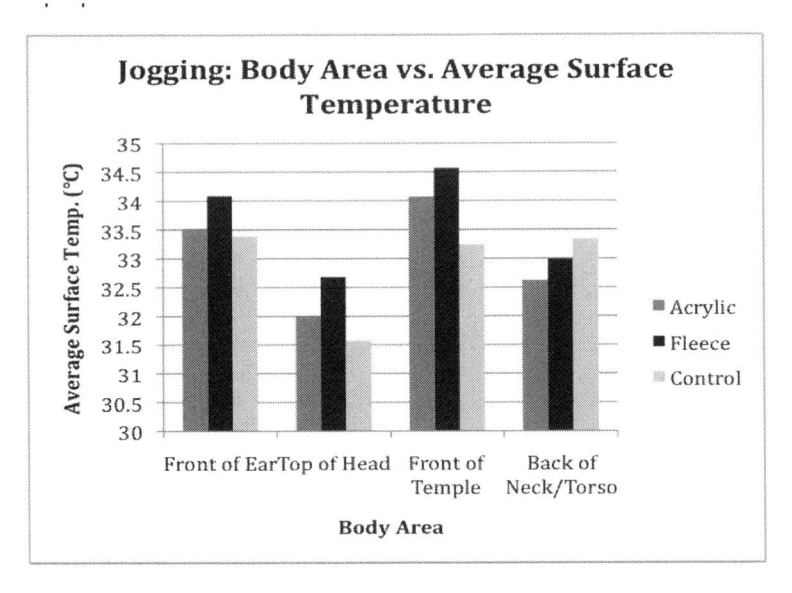

Figure 18. A graphical comparison, between acrylic and fleece hats and the control, of the average body temperatures achieved at various parts of the body while jogging. The fleece hat is shown to result in highest average body temperatures for all measured body areas other than the back of neck/torso.

	Hot Water Temp. (°C)	T-Test

	(Power: Full Flow vs. Crimped Flow)
34	0.3380
37	0.0480
40	0.0770
Control	0.0320
All	0.0011

Table 6. Results of Student T-tests performed to compare the measured output powers of the generator for various hot water temperatures during Experiment 2 under full flow and crimped flow conditions.

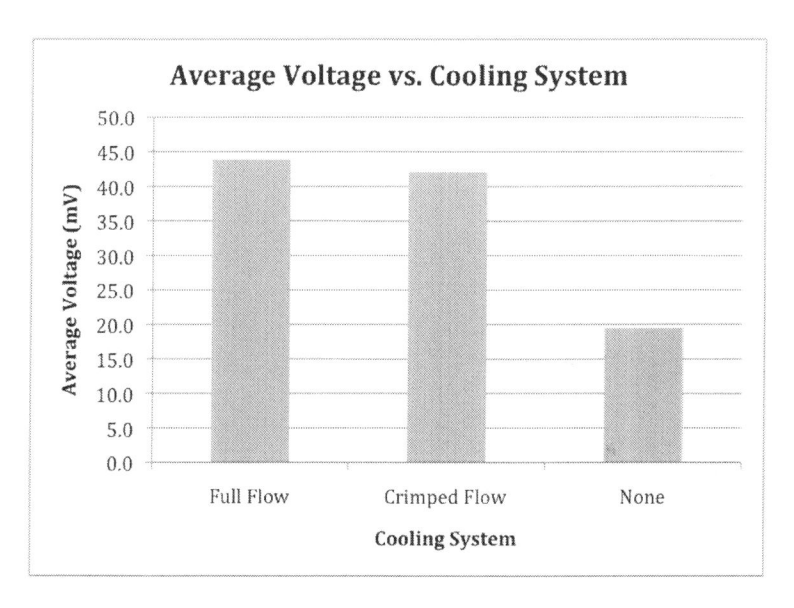

Figure 19. A graphical comparison of the average output voltages of the generator for each of the cooling systems tested in Experiment 2. The full flow system is shown to produce the highest average voltage.

Prototype	STD EV	95% CI	%RSD	T-Test (vs. A)	T-Test (vs. C)
Voltage (mV)					
A	1.605	1.342	4.841	-	< 0.0001
B	1.612	1.347	4.573	0.0271	< 0.0001
C	1.811	1.514	2.251	0.0001	-
Control	1.432	1.197	10.294	< 0.0001	< 0.0001
Current (mA)					
A	0.560	0.468	8.145	-	< 0.0001
B	0.590	0.493	7.685	0.0189	< 0.0001
C	0.431	0.360	4.249	< 0.0001	-

Figure 20: ...ical comparison of the average output currents of ... erator for each of the cooling systems tested in ...nt 2. The full flow system is shown to produce t... st average current.

Figure 21: ...ical comparison of the average output powers of the generator for each of the cooling systems tested in Experiment 2. The full flow system is shown to produce the highest average power.

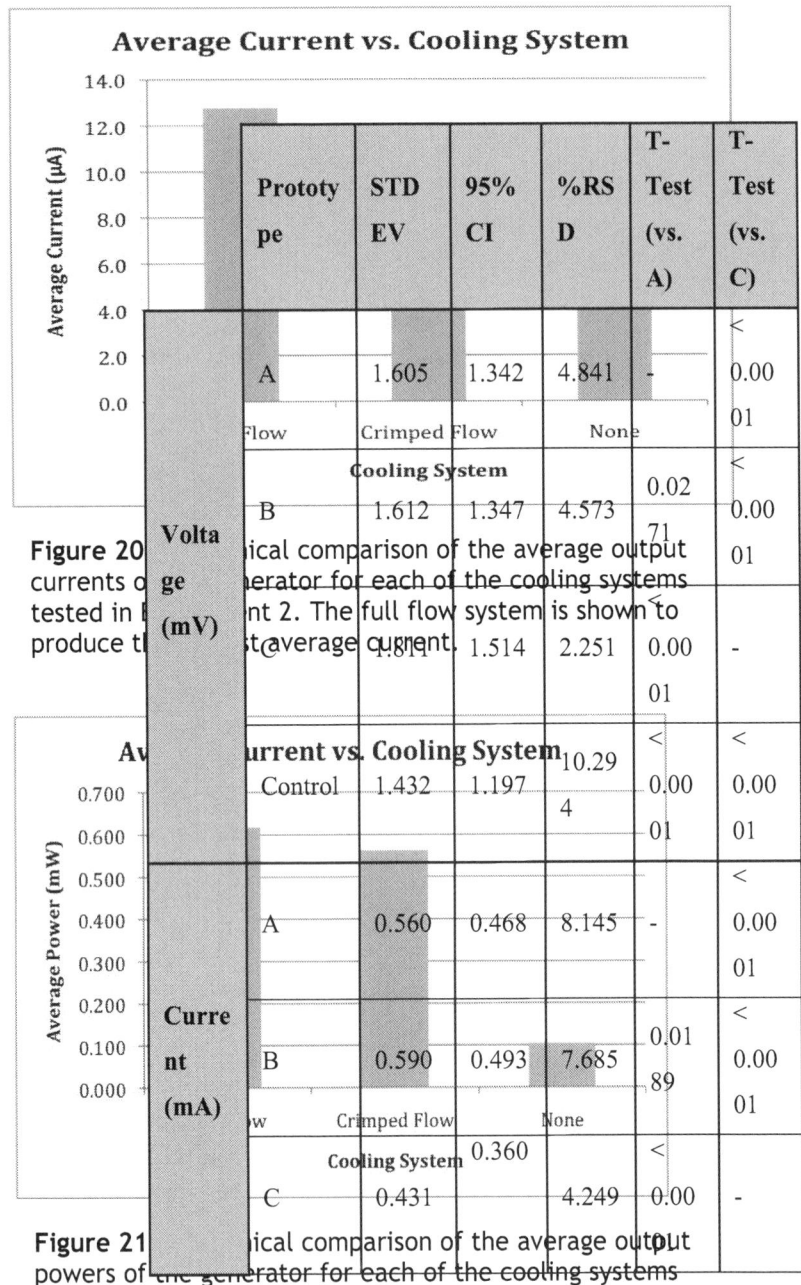

	Control	0.642	0.537	22.364	< 0.0001	< 0.0001
Power (mW)	A	0.029	0.024	12.757	-	< 0.0001
	B	0.033	0.027	11.986	0.0208	< 0.0001
	C	0.049	0.041	6.044	< 0.0001	-
	Control	0.013	0.011	32.200	< 0.0001	< 0.0001

Table 7. Chart of various statistical values gathered from analysis of the voltage, current, and power outputs of the prototype designs from Experiment 3, including standard deviation, percent relative standard deviation, 95% confidence interval, and Student T-test results

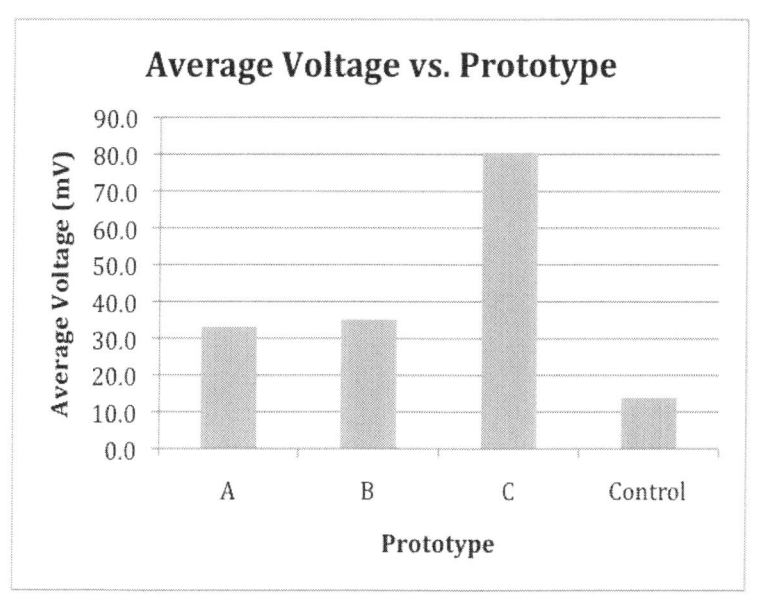

Figure 22. A graphical comparison of the average output voltages of each of the prototypes for Experiment 3. Prototype B is shown to produce the highest average voltage.

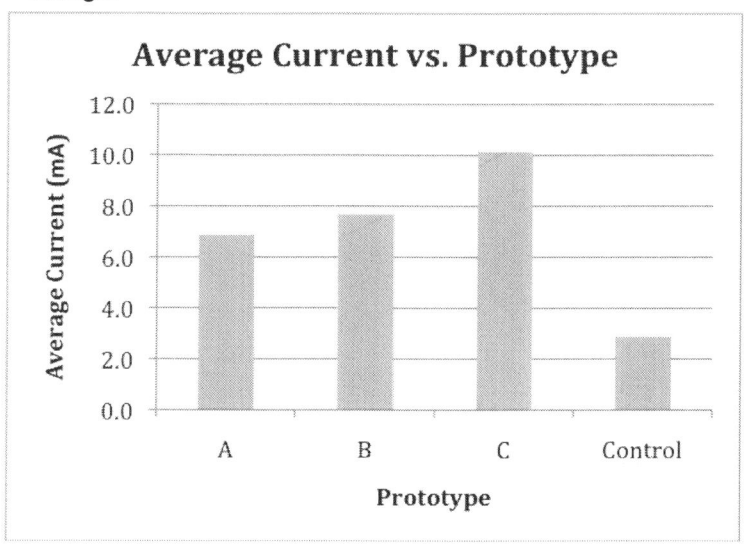

Figure 23. A graphical comparison of the average output currents of each of the prototypes for Experiment 3. Prototype B is shown to produce the highest average current.

Average Power vs. Prototype

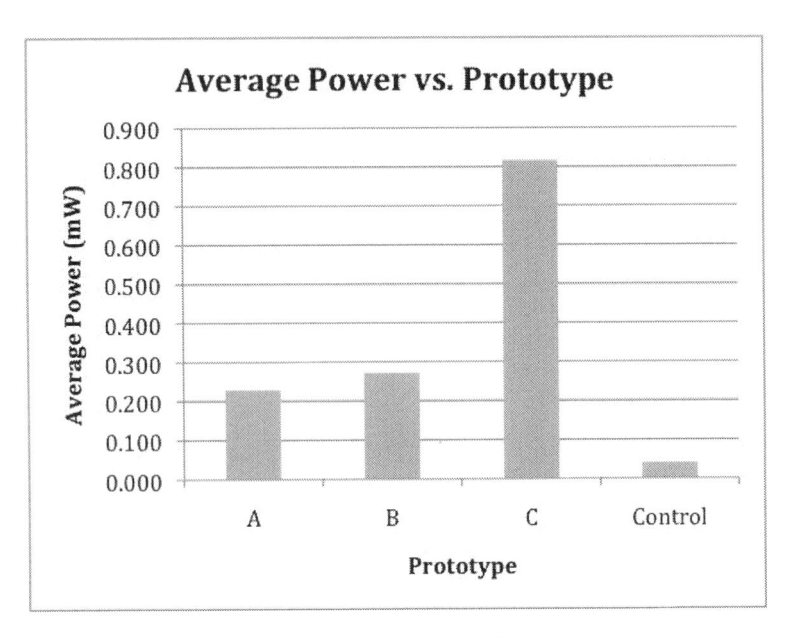

[Bar chart showing Average Power (mW) on y-axis from 0.000 to 0.900, and Prototype on x-axis with categories A, B, C, Control. Bar values approximately: A ≈ 0.23, B ≈ 0.27, C ≈ 0.82, Control ≈ 0.05.]

Figure 24. A graphical comparison of the average output powers of each of the prototypes for Experiment 3. Prototype B is shown to produce the highest average power.

For Experiment 1, Student T-tests performed to compare the performances of the acrylic and fleece hats at insulating the head of the investigator yielded results of less than 0.05 in all cases except for temperature measurements taken in front of the ear during non-active testing and at the back of the neck and torso during testing conducted while jogging. By convention, a student t-test value of less than 0.05 suggests that the null hypothesis, that there is no statistically significant difference between the insulating capabilities of the acrylic and fleece hats, may be refuted with a strong degree of confidence. For those values that were greater than 0.05—0.0957 for the non-active, front of ear measurements and 0.0953 for the jogging, back of neck and torso measurements—the value yielded was less than 0.1, suggesting that the null hypothesis may still be refuted with a moderate degree of confidence. When all of the experimental data are considered, a Student T-test yields a result of 5.38×10^{-13}, a value very near zero and which supports the refutation of the null hypothesis. When additional Student T-tests were performed to compare the body temperatures measured with the fleece hat to the control, all values were $\leq$ 0.05 other than the upper thoracic epidermal measurements taken while jogging. Because the average temperature values measured with the fleece hat were greater than those measured with the acrylic hat in all cases and all control measurements determined through T-testing to be statistically significant, this suggests that the fleece hat is significantly superior to the acrylic hat and the control as an insulating article of headgear. The calculated average epidermal temperature values for the acrylic and fleece hats and control are given in the following table:

Table 8. Average body temperatures for both hat materials and the control for the various body regions, as measured in Experiment 1.

Hat	Activity	Body Area	Average Temp (°C)
Acrylic	None	Front Ear	33.65
		Top Head	31.46
		Temple	33.99
		Neck/Torso	33.48
	Jogging	Front Ear	33.80
		Top Head	32.03
		Temple	34.14

		Neck/Torso	33.87
Fleece	None	Front Ear	33.52
		Top Head	32.01
		Temple	34.07
		Neck/Torso	32.62
	Jogging	Front Ear	34.09
		Top Head	32.68
		Temple	34.57
		Neck/Torso	32.99
Control	None	Front Ear	32.02
		Top Head	31.35
		Temple	32.04
		Neck/Torso	31.49
	Jogging	Front Ear	33.39
		Top Head	31.57
		Temple	33.24
		Neck/Torso	33.34

In Experiment 2, Student T-tests were performed to compare the power outputs of the generator with full flow and crimped flow cooling systems at various hot water temperatures. For a hot water temperature of 34°C, the T-test yielded a value of 0.338, suggesting that the null hypothesis could not be refuted with sufficiently high confidence. For a temperature of 40°C, the test yielded a value of 0.077, suggesting that the null hypothesis could be refuted with moderate confidence.

For the 37°C and control settings, the tests yielded values of less than 0.05, suggesting that the null hypothesis could be refuted with high confidence. Finally, when the data was considered as a whole, a T-test yielded a value of 0.0011, suggesting that the null hypothesis could be refuted with high confidence. Assuming the null hypothesis can be refuted, the average voltage, current, and power output measurements from the experiments suggest that the full flow cooling system yielded the greatest output, but because these values are so close, this data suggests that a slower rate of flow may actually be more practical if the performance of the device does not suffer too significantly. The calculated average voltage, current, and power outputs for each cooling system that was tested are given in the following table.

Cooling System	Voltage (mV)	Current (μA)	Power (mW)
Full Flow	43.9	12.8	0.617
Crimped Flow	42.1	12.0	0.563
No Flow	19.5	4.0	0.104

Table 9. Average voltage current and power outputs of the generator for each of the three cooling systems tested in Experiment 2.

During this testing, the time for the water input reservoir was found to drain within 1.25 minutes with the full flow system and within two minutes with the crimped flow system.

For Experiment 3, the second measured voltage, current, and power of the control setup were determined to be outliers and were not considered in statistical analysis. The percent relative standard deviations (%RSD) for the voltage, current, and power outputs of each prototype and the control were determined. The %RSD values for the voltages measured from Prototypes A, B, and C were determined to be less than five, meaning that these data were measured with high precision. The %RSD values for the current outputs of Prototypes A and B were both determined to be less than 10, meaning that these data were measured with moderate precision, and that of Prototype C was measured to be less than 5. All other %RSD values determined that the remaining data were measured with relatively low precision. Student T-tests performed to compare each of the prototypes and the control yielded values of less than 0.05, meaning that the null hypotheses for the performance of these devices can be refuted with high confidence. Because Prototype C is shown to produce a higher average voltage, current, and output than Prototypes A and B and the control setup in every case, it can be determined with a high degree of confidence that Prototype C performs significantly better than any of the other setups tested in this experiment. The average voltage, current, and power outputs of each of the prototypes and the control setup are shown in the table below:

Prototype	Voltage (mV)	Current (mA)	Power (mW)
A	33.2	6.9	0.229
B	35.2	7.7	0.271
C	80.5	10.1	0.816
Control	13.9	2.9	0.041

Table 10. Average voltage current and power outputs of each of the prototypes and control setup tested in Experiment 3.

During this testing, the input reservoir for Prototype B was found to drain within three minutes, a longer duration than was measured for the setup tested in Experiment 2. That of Prototype C was found to drain within four minutes.

Conclusion

Based on the results gathered in this investigation, the final prototype design does not fulfill the engineering goal of generating electricity sufficient to power any common electrical device other than a simple capacitor circuit concurrently to the use of the device. Furthermore, because of its weight, component cost, and brief operating period, everyday use of the design would not be practical, and it cannot currently stand on its own as a viable device for market distribution. However, the data gathered show that the design of Prototype C was a significant improvement over those of Prototypes A and B. The data also provide insight into the operation of such a device and is likely to aid in the assistance of future iterations of prototypes of this device. For example, a sufficient number of these devices wired in parallel over the various exposed surfaces of the human body could potentially provide sufficient voltage and current to achieve the engineering goal, and the prototype design, with a backpack acting as an isolated containment unit, allows for the incorporation of any conversion circuits that may need to be attached to the output electrical wiring in order to harness this method of thermoelectric conversion for more viable and directly practical applications.

Limitations and Assumptions

The final prototype possesses several limitations as a practical product and as a viable alternative energy source. Because of restrictions related to available budget, time, and other resources, only a single generator and cooling system design could be physically constructed and subsequently tested. The voltage and current outputs provided by this single generator when incorporated into each prototype design are not sufficient to consistently power a handheld device without transformation, storage, and subsequent discharge. In order to provide sufficient power to accomplish this, more generators are needed; however, because of the weight and size of the current generator design and the limited space offered by the exposed cranial and thoracic surface area of an individual, this could potentially render the headgear to which they are affixed impractical as an article of clothing. Also, the current design requires an unrestricted flow of ice water through the generator in order to operate at its measured level of performance; however, this causes the input water source to be depleted quite rapidly, within approximately four minutes. In order for the device to be practically suitable for use as a portable generator for a sustained period of time, it would need to operate for a much longer duration without requiring the maintenance of the user.

Additionally, several assumptions were made in order to complete the necessary processes of design and experimentation within the time, budget, and resources available. First, it was assumed that the device would be utilized in a controlled, room temperature environment of approximately 20°C. Also, the assumption was made that any internal resistance offered by the thermoelectric generator was negligible. Furthermore, a rubber hot water bottle was assumed to transfer heat to the thermal junction of the generator in a manner identical to human skin tissue. The bodily characteristics and behaviors of the investigator, including height and metabolic heat generation, were also assumed to serve as an accurate model for those of the average general population.

During experimentation, multiple variables were controlled in order to minimize their potential effect on experimental results. For body temperature testing, the hat was not altered or replaced at any time during experimentation and was left at room temperature prior to the start of the experiment. Also, no food was consumed by the investigator for at least four hours prior to testing to prevent unanticipated effects to metabolic processes, and no sources of radiant heat were present during experimentation; however, small fluctuations in temperature between the five minute intervals at which it was measured could not be controlled entirely. For generator and prototype performance testing, the generator was not altered or replaced at any time during experimentation and was left at room temperature prior to the start of experimentation. Although the surface temperature of the hot water bottle was measured during this phase of experimentation, the gradual normalization of the temperature of the water in the system because of the time required to record measurements could not be controlled and may have led to some inaccuracy in the data.

Due to the nature of testing, several sources of error were present in the collection of data. One potential source of error may have been human imprecision in reading measurement devices such as the alcohol thermometer used in experimentation. Further errors in measurement may have resulted from the non-ideal efficiency of the circuits and electrical components used in experimentation.

Applications and Future Extensions

The engineered prototype allows thermoelectric harvesting technology to draw upon the renewable waste heat generated by the cranial and thoracic regions of the human body as an energy source to provide a usable electric output energy to charge or power electrical devices. The device therefore serves as an archetype for utilizing any small, renewable source of heat to provide usable electric energy to a battery or a variety of other common devices. While most previous applications of thermoelectric technology have relied upon a relatively abundant source of heat to produce a usable amount of output energy, the technology and design implemented in this device serve as a model for how this technology may be applied on a smaller, more reasonably maintainable, and more practical scale.

Further research into improving upon the design proposed in this investigation may involve redesigning the prototype to incorporate an array of generators wired in series to one another in order to increase the voltage output of the apparatus and thus the current output for an identical load. The amount of voltage produced could then be stepped up or down as needed using a DC-to-DC transformer. The voltage produced could be provided to a capacitor for storage and eventual discharge to an attached load in the form of a battery or other electrical device.

Alternatively, future research may encompass further increasing the efficiency of the heat transfer from the epidermis to the external thermal junction of the generator. This would minimize the need for multiple generators to be wired in series to provide sufficient voltage for DC-to-DC conversion, thereby removing the possibility of inefficiency in series wiring and rendering the device more lightweight and practical. This could potentially be accomplished by utilizing alternatives to the semiconductive bismuth telluride material that constitutes a component of the thermoelectric generator circuit. The same effect could be achieved by increasing the surface area of the external junction in contact with the cranial epidermal surface of the user, possibly by contouring the plate to fit the curved shape of the crown of the head. Future technology could also be employed to render the generator more lightweight and thus more practical as a device for consumers who will be expected to wear it as they would an article of clothing.

Literature Cited

Angrist, S. W. (1982). *Direct energy conversion* (4th ed.). Boston: Allyn and Bacon.

Body heat to power cell phones? Nanowires enable recovery of waste heat energy. (2008). Retrieved November 15, 2010, from *ScienceDaily*: http://www.sciencedaily.com/releases/2008/01/080110161823.htm

Brain M. and Bryant C. W. (2000). *How batteries work*. Retrieved November 25, 2010, from http://electronics.howstuffworks.com/battery.htm

Car battery charger circuit. (2010). Retrieved November 15, 2010, from http://www.free-circuit.com/wp-content/uploads/2010/04/car-battary-charger-circuit.gif

Carmen, R. (ed.). (2004). *Hearing loss and hearing aids: a bridge to healing.* 2nd ed. Sedona, Arizona: Auricle Ink.

Cross, A., Collard M., and Nelson A. (2008). Body segment differences in surface area, skin temperature and 3d displacement and the estimation of heat balance during locomotion in hominins. *PLoS ONE*, 3 (6): e2464, 1-9.

DC-DC Converter Tutorial (2001). Retrieved February 16, 2011, from http://www. maxim-ic.com/app-notes/index.mvp/id/2031

Decher, R. (1997). *Direct energy conversion: fundamentals of electric power production.* New York: Oxford University Press.

Dillner, U. (2010). Can thermotunneling improve the currently realized thermoelectric conversion efficiency? *Journal of Electronic Materials*, 35, 1645-1649.

Fleming, J. W. (2010). Ultra-low power conversion and management techniques for thermoelectric energy harvesting applications. *Proceedings of SPIE,* 7683.

Gibilisco, S. (2001). *Teach yourself electricity and electronics*. 3rd ed. New York: McGraw-Hill.

Kaminski, D. A. and Jensen, M. K. (2005). *Introduction to thermal and fluids engineering.* Hoboken, New Jersey: John Wiley & Sons.

Kightly, R. n.d. *Video animation: simple electrical circuit showing current flow.* Retrieved November 8, 2010, from http://www.rkm.com.au/ANIMATIONS/ animation-electrical-circuit.html

Layton, J. (2006). *How iPods work.* Retrieved November 25, 2010, from http:// electronics.howstuffworks.com/ipod.htm

Pawelski, J. (2006). *Next generation thermo-electric systems - family of projects.* Retrieved November 15, 2010, from http://edge.rit.edu/content/P07440/ public/Home

Rowe, D. M. (1999). Thermoelectrics, an environmentally-friendly source of electrical power. *Renewable Energy*, 16, 1251-1256.

Shitzer, A. and Eberhart, R. C. (eds.). (1985a). *Heat transfer in medicine and biology: analysis and applications* (Vol. I). New York: Plenum Press.

Shitzer, A. and Eberhart, R. C. (eds.). (1985b). *Heat transfer in medicine and biology: analysis and applications* (Vol. II). New York: Plenum Press.

Simple boost converter (2001). Retrieved February 16, 2011, from http://www.maxim-ic.com/app-notes/index.mvp/id/2031

Spakovszky, Z. S. (2002). *Unified: thermodynamics and propulsion.* Retrieved November 25, 2010, from http://web.mit.edu/16.unified/www/FALL/thermodynamics/notes/notes.html

Strachan, J. S. and Aspden, H. (1994). U.S. Patent No. 5,288,336. Washington D.C.: U.S. Patent and Trademark Office.

Su J., Vullers R., Goedbloed M., Andel Y., Leonov V., and Wang Z. (2010). Thermoelectric energy harvester fabricated by Stepper. *Microelectronic Engineering*, 87, 1242-1244.

Thermodynamics, heat transfer, and fluid flow. (1992). Vol. 1. Washington, D.C.: U.S. Department of Energy.

Yamashita, O. (2009). Effect of linear and non-linear components in the temperature dependences of thermoelectric properties on the energy conversion efficiency. *Energy Conversion and Management*, 50, 1968-1975.

Effect of Sounds on Marine Mammals
By Carlo Bocconcelli

Abstract

Studies on the effect of sonar, and other sounds, on marine mammals are very important for the understanding of interactions between precarious populations of marine mammals and the increasingly commercialized ocean. Sonar is used for military exercises and also by ships, and there are few regulations on its usage and maximum loudness. If sonar was to potentially startle or injure marine mammals, or deter them from breeding migrations or communication, new laws that limit the use and potency of sonar must be considered. Many whale species are already endangered because of habitat destruction and human interactions. This experiment assessed the calls produced by two pilot whales recorded with DTAGS by the Woods Hole Oceanographic Institution in 2007 using specially created MATLAB programs and input into an Excel spreadsheet. The project looked at the call times and call types recorded on the spreadsheet and whether the overall call rates and stereotyped (distinct, repeated calls) call rates changed as the recorded pilot whales were subjected to sonar and killer whale sounds. The results of this project suggest that sonar pulses ranging from 150-200 dB of loudness caused pilot whales to produce averages of around 5 calls/minute less than their original calls/minute, while killer whale playbacks made the same pilot whales increase their vocalizations, but not to the original levels. The project indicates that sonar makes whales vocalize less, and therefore indicates that projects like this should be repeated in the future with the pilot whales, and new species, to determine the effects of sonar on marine mammals and to protect them from the potential dangers of sonar.

Background

Over the past decade, increasing attention has been drawn to the impact of human-generated sound on whales. The study of pilot whales is especially important because of the pilot whales' tendency for mass stranding. In some of these cases, the strandings were thought to be the consequence of man-made sound (Sayigh, 2010). One of the most intrusive man-made sounds comes from active sonar used by navy and commercial ships. Sonar is a way to detect submersed objects near the emitter of the sonar wave. Sonar is used by emitting a sound wave that travels through the medium (water) until it reaches another object, which it bounces off, returning to the emitter (as an echo). The emitter of the sound wave uses the time elapsed between the departure and arrival of the sound wave to calculate the distance from itself to the target object (Nicholson, 2008, Internet). Navy vessels use particularly loud sound waves, which travel further. However, these loud, intrusive sound waves can harm marine mammals near them. Marine mammals use their own form of sonar, termed echolocation, to find prey.

Because of this, their hearing is developed and sensitive to receive the echoes of their own calls and therefore loud Navy sonars could damage their brain and ears. Sonar also scares these mammals away from their foraging areas and can cause them to surface quickly, causing decompression sickness, commonly called the bends (Nicholson, 2008, Internet).

Pilot whale vocalizations in response to sonar and killer whale sounds will be analyzed in this experiment to determine the effects of these sounds on their vocal behavior. Pilot whales are in the class Mammalia, order Cetacea, suborder Odontoceti, family Delphinidae, genus *Globicephala* and species *melaena* (long-finned) and *macrorhynchus* (short-finned). This experiment will be analyzing the calls of short-finned pilot whales, so most of the information on pilot whales will be about short-fins (American Cetacean Society 2006, Internet). Long-finned pilot whales have large flippers (1/5 of their body length) and are found in cold areas, whereas short-finned pilot whales have shorter flippers (1/6 of their body length) and are found in warmer areas (Martin, 1990). Also, most short-finned pilot whales have unique dorsal fins, which make individual whales identifiable. Pilot whales have a bulbous head, which houses their melon, the organ that aids their use of echolocation. Pilot whales range from jet black to dark gray. Adult males may be heavily scarred from fights with other males over females. Their dorsal fin shape varies according to sex. Pilot whales have very robust bodies and a thick tail stock, which allows them to achieve very fast swimming speeds. They have long, slender sickle shaped flippers positioned very close to their head. Adults are 12-21½ feet long, with males larger than females, and they weigh 1-4 tons. Short-finned pilot whale group sizes range from 1-50 (Cardawine, 2000).

Pilot whales are closely related to the killer whale, *Orcinus orca,* which is also their main predator. This experiment will also analyze the effects of killer whale sounds on pilot whale vocal behavior, as killer whale vocalizations have been known to sound much like sonar. Pilot whales can be similar to killer whales socially, but killer whales have many physical differences (American Cetacean Society 2006, Internet). Killer whales grow 5.7 to 8 meters in length; have a tall dorsal fin and a powerful body. Killer whales have a notched tail and paddle like flippers, enabling them to swim very fast. Killer whales usually travel in pods of 3-25 animals with at least one large male. They can tolerate any water conditions, allowing them journey to and hunt in a variety of food-rich regions. They have tight social bonds and hunt cooperatively, allowing them to take down any size prey. Killer whales breed inside of their pods and are not in danger of extinction, but could lose much of their prey from overfishing.

Short finned pilot whales eat squid and are found mainly in warm waters (Martin, 1990). Short-fins feed at night with dives of 10 minutes or more, and prefer deep water (Cardawine, 2000). Short-fins have several distinct populations and are sometimes nomadic, following their prey. Their groups are smaller than those of the long-finned pilot whale. The two species of pilot whale generally do not overlap much due to their temperature constraints (warm vs. cold water), therefore reducing crossbreeding between the two species (Martin, 1990). Short-finned pilot whales often do something called logging, in which the animal rests and floats at the surface of the water. Short-finned pilot whale pods are very social, and have strong matrilineal associations (Cardawine, 2000). These whales have a breeding interval of 3-9 years. A distinctive characteristic of all pilot whales is that they may reach reproductive senescence. The older females in the pod who have reached reproductive senescence switch their goals from calf bearing to calf rearing. They can raise and take care of the calves in the pod while the calves' mothers go hunting, increasing social stability. Most female short-finned pilot whales have a post-reproductive lifespan of 20-30 years. The calf gestation period lasts 12-16 months, and the lactation period lasts for about 42 months. Since short-finned pilot whales spend so much time around individuals related to them, their social bonds are further increased. This also increases their hunting coordination and cohesion. Short-finned pilot whales can live up to 60 years. The adult mortality rate is 4% more in males than in females, and juveniles have a 10% mortality rate. Males reach sexual maturity at 16 years, and females at 8-9 years. They breed with distantly related pods to avoid inbreeding (Evans, 2001).

Short-finned pilot whales are still killed for their oil and meat, and as a result their overall population is reduced. Because of this, two genetically distinct populations, which lived mainly off Japan, have been hunted to extinction (Cardawine, 2000). They can also get entangled in fishing gear, which is an issue of increasing concern as commercial fisheries become more abundant (Martin, 1990). Pilot whales have also been caught in long lines, drift nets and gill nets after they try to take fish from them (Evans, 2001). When entire pods log, which they are known to do, they are sometimes hit by boats. They are generally indifferent to boats, which increases their chance of being suddenly struck by a fast-moving boat (Cardawine, 2000). Pilot whales are very prone to mineral deficiencies and health impairments caused by loss of prey, which can reduce breeding (Evans, 2001).

Pilot whales are very intelligent and have very strong social bonds. Because of this, pilot whales are prone to mass stranding, as one sick whale can lead the others to strand (American Cetacean Society, Internet). The reason for the healthy ones following the injured one is so far unknown (Cardawine, 2000). Short-finned pilot whales strand in groups of 20-90 (Martin, 1990).

Before talking about whale echolocation, and to further explain sonar, it is necessary to give a basic introduction to the properties of sound. A sound wave transports energy (and information) through a medium, which can be a liquid, such as water, a gas, such as air, or a solid. Sound waves do not change the medium itself, but create a disturbance. This disturbance briefly changes the density of the medium, and this change can be detected to interpret the wave. Sound waves are measured in Hertz (Hz), which is a unit of frequency; a periodic interval of one second (units of S^{-1}). One Hz is a complete cycle of a wave, and a KHz is a unit equal to 1,000 Hz. A decibel (dB) is a logarithmic unit of sound intensity, equaling 10 times the logarithm of the ratio of the sound intensity to some reference intensity. Sound in water will have a larger dB value than the same sound in air because of different reference levels and the differences in the density and compressibility of air and water (Rossing, 1990). Sound is very important to pilot whales, as they use it for communication and hunting.

Pilot whales, killer whales and all other toothed whales (odontoceti) use echolocation, or biosonar, to find prey. High-frequency clicks are projected narrowly forward (from the head of the whale) when hunting. Clicks are produced by the animal and received back as echoes. From the echoes, the animal determines the size and shape of the object the click bounced off of and the distance from the object. To create clicks, sound is emitted from the nasal passages of the animal, focused through the melon (a fatty area of the head) and projected from the animal. The echoes are received by the inner ear after they pass through sound-conducting tissue in the lower jaw. Whales accelerate clicks to a buzz when closing in on their prey (Tyack, 2009). Killer whales use a variety of vocalizations: short broad-band pulses of 0.8-25 milliseconds, used for hunting echolocation, tonal signals and whistles of 1.5-18 KHz lasting 0.05-12 seconds, and burst pulses of up to 25 KHz lasting 0.5-1.5 seconds (Martin, 1990). Pilot whales produce similar types of vocalizations, although theirs are not as well studied. Long-finned pilot whale calls have a more narrow range and are lower in frequency than those of the short-finned pilot whale. Long-finned pilot whale whistles range from 1-8 KHz, while short-finned pilot whale vocalizations range from 2-14 KHz. The average duration is 0.6 seconds. Pilot whales have been found to produce repeated stereotyped calls that often occur close to one another. There are many distinct types of calls, but it is not known whether each call type is produced by only one or by more than one whale.

There is evidence that pilot whales and other species of whales may strand as a result of sonar, illustrated in these incidents. Recently, 33 pilot whales mass stranded off Rutland Island in Ireland. The cause of death is unsure, but it could have been a storm, which drove the pod close to shore, or hearing trauma of certain whales caused by Royal Navy sonar (McCormack, Internet). In 2007, five beached whales were found in California. The cause of death was determined to be sonar from a naval exercise nearby, and this led to a two-year ban on sonar off the coast of California (McCormack, Internet). In the year 2000, U.S. Navy destroyers used sonar during a training exercise in the Bahamas. Within 36 hours of the exercise 17 whales stranded on the beaches in the surrounding areas with ear and brain traumas that most likely caused swimming and navigational issues. It is believed that mass strandings occur routinely during Navy sonar exercises (Nevala, 2009). Areas with steep ocean floor cliffs and varied temperatures can cause sonar sound waves to travel faster, and these are the areas where most mass strandings happen. Different intensities of sonar may also replicate sounds made by other whales, such as killer whales, which may frighten other whales and cause them to act abnormally (Nevala, 2009). It has been found that these whales will ascend too quickly in an effort to escape, resulting in them getting the bends. This quick change in pressure, caused by surfacing too suddenly, causes the diffused gases in the blood stream and tissues to expand and injure the nervous and audiovestibular (inner ear) systems (Madsen, 2010). A previous study showed that when beaked whales (*Ziphiidae*) are exposed to experimental playbacks of sonar, they quieted and left the area. Pilot whales exposed to the same signals appeared to increase vocalizations and form cohesive groups (Nevala, 2009). This research indicates that pilot whales will sometimes increase vocalizing and social cohesion when exposed to sonar and killer whale noises, but sometimes they will leave the area (Tyack, 2009).

There are few data available on how whales respond to sonar. In November 2008, the U.S. Supreme Court ruled in favor of the Navy in the Whales vs. Sonar case. Animal activists wanted the shutdown range of sonar to be increased from the current distance of 1,000 meters from a marine mammal to 2000 meters. Since there is little proof that stranded whales suffered decompression sickness from surfacing quickly after being startled by sonar, the Supreme Court decided there was not enough evidence to tighten sonar regulations. However, the Navy now funds many research operations to discover how marine mammals are affected by sonar (Madin, 2009).

Sonar is only one of the issues posed by sound disturbance, as noise caused by boats and machinery interferes with the lives of marine mammals. Boat noises affect the transmission of whale vocalizations, reducing breeding call strength in many species, which eventually could reduce the overall population due to reduced breeding. Ambient noise forces whales to expend more energy just to make their signals heard. Whales can be diverted from their migrations by loud noises, which scare them from their usual routes. High decibel sounds damage the eardrums and lungs of marine mammals. U.S. regulations establish 160 dB as the disruption threshold for whales and dolphins (Tyack, 2009).

Much of the ongoing research about the effects of sonar on whales uses Dtags to collect data on whales. Dtags, or digital acoustic tags, are placed on whales by means of suction cups. These tags record sounds produced and received, and also record depth, pitch, roll and heading (movements and direction of the whale). Dtags are vital in determining the effects of sonar on whales, as they can record the whales' responses to different sounds (Madin, 2009). Dtags use three-axis magnetometers and accelerometers, pressure sensors and hydrophones, and can remain on the animal for many hours, providing extensive data. Recent experiments have exposed pilot whales, wearing Dtags, to playbacks of sonar and killer whale sounds to determine their reactions to these sounds (Tyack, 2009). For this experiment, the information recorded by Dtags placed on short-finned pilot whales subjected to sonar and killer whale playbacks will be analyzed, and the vocalizations produced by the whales will be examined to understand how these whales respond to said playbacks.

Hypothesis

It was hypothesized that initially the pilot whales would be vocalizing, that during the sonar and killer whale sounds the pilot whales would stop vocalizing (and swim away) and that later their vocalization rates would recover to before-playback levels. It was also hypothesized that the stereotyped call trends would follow those of the overall call average trends.

Methods

- Data were gathered in the Bahamas by Woods Hole Oceanographic Institution (WHOI) scientists in

August of 2007 and were analyzed using a WHOI Dell laptop. The data consisted of recordings of the vocalizations of two tagged pilot whales, which were subjected to playbacks of sonar and killer whale sounds. The whales were tagged with Digital Acoustic Tags (DTAGS), which were placed on the animals by a research team led by Dr. Peter Tyack at WHOI.

- Data were obtained from WHOI scientist Dr. Laela Sayigh on a Verbatim 500gb flash drive.
- Notebook containing times of sonar and killer whale playbacks played to the two tagged pilot whales (one was tagged by DtagA, the other by DtagB) was obtained from Dr. Sayigh.
- Matlab programs were used to verify the occurrences and times of sonar and killer whale sounds in the pilot whale recordings contained on the Verbatim flash drive.
- The Excel file titled TagA, which was on the Verbatim flash drive and contained all the times of the sounds recorded by DtagA, was opened on the Dell laptop, and the times at which the pilot whale was exposed to sonar and killer whale sounds were entered into an Excel spreadsheet.
- The same procedure was carried out on a second Excel file, titled TagB, which contained all the sounds recorded by DtagB.

- Calls were counted and marked in 10 min intervals on Excels files TagA and TagB until 30 minutes before and after sonar and killer whale playbacks. From 30 minutes before to 30 minutes after playbacks calls were counted in 1-minute intervals to obtain more exact data.
- The calls per minute pre, during, and after sonar, and pre, during and after killer whale sounds on file TagA and file TagB were averaged.
- Averages were graphed using Kaleidagraph.
- Trends in call rates pre, during and after playbacks were examined to see whether sonar or killer whale sounds affected the vocalization rates of pilot whales.

Results

For the first sonar exposure, TagA recorded before, during and after averages of 4.57, 2.80 and 1.46 calls per minute, respectively. Tag B recorded before, during and after averages of 2.74, 7.00 and 2.12 calls per minute, respectively. For the second exposure, TagA recorded before, during and after averages of 1.46, 0.08 and 0.39 calls per minute, respectively. TagB recorded before, during and after averages of 2.12, 0.67 and 0 calls per minute, respectively. For the killer whale call exposure, TagA recorded before, during and after averages of 0.39, 0.21 and 2.67 calls per minute, respectively. TagB recorded before, during and after averages of 0, 2 and 3.38 calls per minute, respectively. Stereotyped calls/minute were also analyzed. For the first sonar exposure, TagA recorded before, during and after averages of 0.914, 0 and 0.805 stereotyped calls per minute, respectively. Tag B recorded before, during and after averages of 2.079, 0 and 1.0 stereotyped call(s) per minute, respectively. For the second exposure, TagA recorded before, during and after averages of 0.805, 0 and 0.321 stereotyped calls per minute, respectively. Tag B recorded before, during and after averages of 1.0, 0.308 and 0 stereotyped call(s) per minute, respectively. For the killer whale sound exposure, TagA recorded before, during and after averages of 0.321, 0.214 and 0.669 stereotyped calls per minute, respectively. TagB recorded before, during and after averages of 0, 0.5 and 1.292 stereotyped calls per minute, respectively.

Table 1: Duration and Content of Sonar and Killer Whale Playback Stages

	TagA1 Sonar	TagB1 Sonar	TagA2 Sonar	TagB2 Sonar	TagA Killer	Tag B Killer
Before	93 min. / 425 calls	38 min. / 104 calls	48 min. / 70 calls	50 min. / 106 calls	33 min. / 13 calls	42 min. / 0 calls

During	5 min. / 14 calls	2 min. / 14 calls	12 min. / 1 call	12 min. / 8 calls	14 min. / 3 calls	15 min. / 30 calls
After	48 min. / 70 calls	50 min. / 106 calls	33 min. / 13 calls	42 min. / 0 calls	509 min. / 1357 calls	24 min. / 81 calls

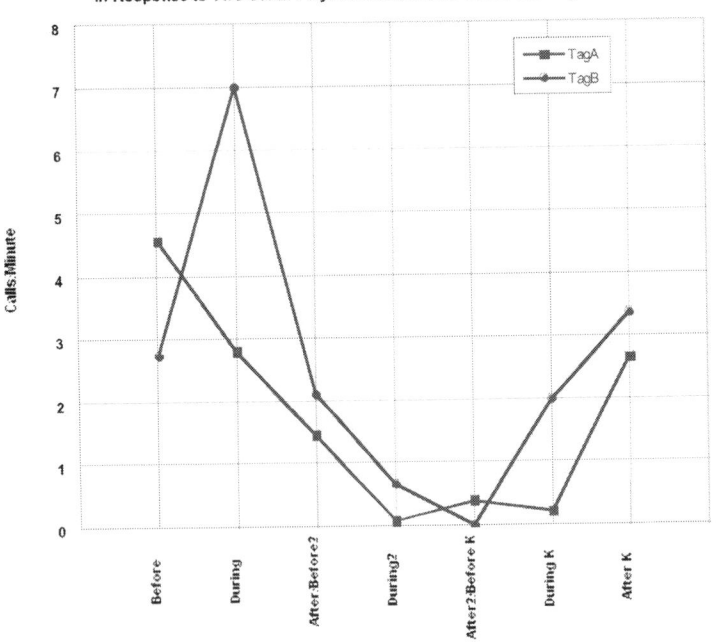

Graph 1: Average Call Rates of Pilot Whales Recorded by TagA and TagB in Response to Two Sonar Playbacks and a Killer Whale Call Playback

Graph 2: Average Stereotyped Call Rates of Pilot Whales Recorded by TagA and TagB in Response to Two Sonar Playbacks and a Killer Whale Call Playback

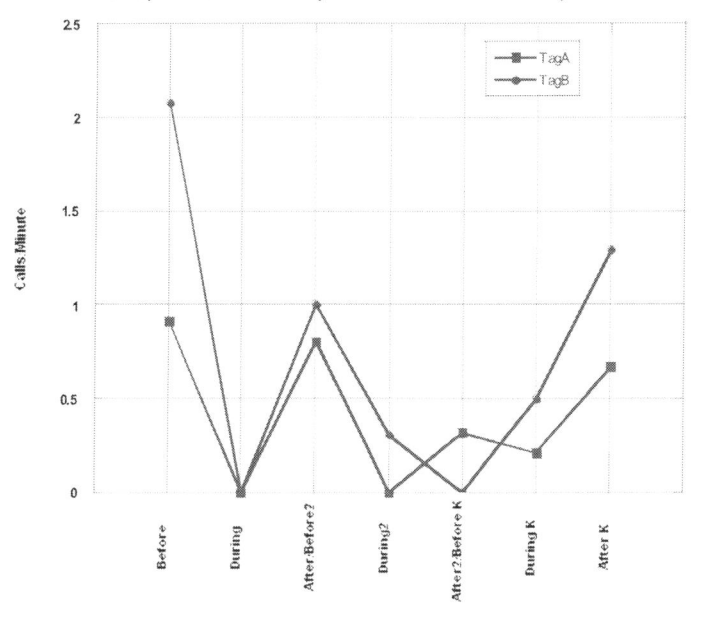

Analysis and Conclusion

The purpose of this experiment was to determine whether short-finned pilot whales would change their average vocalization rates in response to sonar or killer whale sounds, and whether these whales would change their rates of stereotyped calls (distinct repeated calls) in response to sonar or killer whale sounds. It was hypothesized that initially the pilot whales would be vocalizing, that during the sonar and killer whale sounds the pilot whales would stop vocalizing (and swim away) and that later their vocalization rates would recover to before-playback levels. It was also hypothesized that the stereotyped call trends would follow those of the overall call average trends.

Two simultaneously tagged short-finned pilot whales (whale A was recorded by TagA and whale B by TagB) were each exposed to the same two playbacks of sonar. The first sonar sequence ranged from 152 dB to 167 dB, while the second sonar sequence was much louder, ranging from 167 dB to 212 dB. The killer whale playback, exposed to both pilot whales simultaneously, ranged from around 100 dB to 196 dB. The durations of each stage (before, during and after sonar and killer whale playbacks) are shown in Table 1. Also shown in Table 1 are the total numbers of calls that occurred during these stages. The average numbers of calls/minute are plotted in Graph 1. The average numbers of stereotyped calls/minute are displayed in Graph 2. The calls recorded by Tags A and B are not only calls made by the tagged whale; they are also calls made by any whales in close proximity to the tagged whale. However, these whales were also exposed to the same sonars and killer whale calls.

Graph 1 displays average call rates for both whales before, during and after two sonar exposures and a killer whale sound exposure. However, the after period for the first sonar exposure is the same as the before period for the second sonar exposure, and the after period for the second sonar exposure is the same as the before period for the killer whale exposure.

Graph 2 displays average stereotyped call rates for both whales before, during and after two sonar exposures and a killer whale sound exposure (the same exposures as the first graph). Once again, the after period for the first sonar exposure is the same as the before period for the second sonar exposure, and the after period for the second sonar exposure is the same as the before period for the killer whale exposure.

The difference in average calls per minute for both graphs between the first stage of sonar exposure and the second stage of sonar exposure may have occurred for two reasons. One, the first exposure was not as loud as the second exposure and the pilot whales could have confused it for another whale's call. Also, the whales may have been startled by the first exposure so their vocalizations were less in the second exposure stage. The average calls per minute were much less in the second stage of sonar exposure. The second sonar was very loud, intrusive and unnatural, with reverberations that may have startled and confused whales close by. It is possible that the vocalizations during sonar were few because the whales may have quickly gathered back into a group and possibly swam away, and then started vocalizing after they had reached a spot farther away, which supports the theory that exposure to startling sounds may increase social cohesion in pilot whales (Tyack, 2009).

The data trends for TagA and TagB displayed in Graph 1 are closely correlated, except for during the first sonar exposure, when TagA's calls/minute decreased, but TagB's increased to 7 calls/minute (the highest average recorded). This may have occurred because the whale recorded by TagB was at a different distance from the playback source than the TagA whale, or that the whale recorded by TagB was near many other whales whose vocalizations were also picked up by TagB. Aside from this one outlier, the trends remain closely correlated; TagB's recorded calls/minute stay closely above those of TagA, or very closely under. Both tags record a significantly lower number of calls/minute in the second sonar exposure stage, and then increase in the killer whale exposure stage to averages intermediate between the during stages of the first and second exposures, supporting the interpretation that lower dB sonars can sound like a whale's call, while louder sonars and killer whale sounds scare other whales (Tyack, 2009).

The data trends for rates of stereotyped calls on TagA and TagB (displayed in Graph 2) are very closely correlated, though the whales recorded by TagB produced higher averages of stereotyped calls/minute than those recorded by TagA (except for the period after the second sonar exposure, when TagA recorded slightly more stereotyped calls/minute than TagB). Both whales started with high stereotyped calls/minute averages; in the stages after that no such amount was achieved. Both whales dropped to zero stereotyped calls/minute during the first sonar exposure, and during the second exposure TagA once again dropped to zero stereotyped calls/minute, while TagB dropped to less than 0.5 stereotyped calls/minute. After the second sonar exposure, both tags recorded close to zero stereotyped calls/minute, and increased their production in the killer whale playback phase. It would seem that, in the case of stereotyped calls, man-made sounds create a more profound effect than natural sounds, such as killer whale calls. Man-made sounds tend to be louder and reverberate more than sounds made by marine mammals, and they do not contain the same tonal calls and pulses that another whale would recognize (Nevala, 2009). However, since there is not much research on stereotyped calls, their purpose cannot be known for sure (Tyack, 2009).

Though the trends in Graphs 1 and 2 are correlated, both whales showed a greater reduction in stereotyped call rates during sonar exposures, which may indicate that stereotyped calls serve different purposes than normal calls. Also, the whales recorded by TagB almost always produced a larger average of calls, stereotyped or not, than those recorded by Tag A. Overall call rates decreased to almost zero during and after the second exposure in Graph 1. This supports the background data that military sonar startles, disorients and possibly scares away whales (Nicholson, Internet). It is important to notice that for both tags, the stage after the first sonar exposure and before the second exposure are the same, as the two playbacks of sonar were played relatively close to each other in time. Likewise, the stage after the second sonar exposure is also the same as the stage before the killer whale playback. Also, the whales may have been already startled by the second sonar exposure; therefore, their vocalization rates in the killer whale playback stage may have been different from an initial reaction (not having been exposed to sonar beforehand). The same possibility is true for the first sonar exposure influencing the whales' reactions to the second exposure.

An important observation to note is that the before, during and after lengths of all three exposures vary between tags A and B (Table 1). The during stage of the first sonar exposure in TagB has the same amount of calls as in that of TagA; however, the TagA during stage is 3 minutes longer, which accounts for the difference in average calls/minute. These discrepancies can be attributed to a difference in sound pickup from TagA to TagB (due to varying distances from the playback source), or errors in record keeping or data analysis from the tags. The TagB recording also started 55 minutes later and ended 485 minutes earlier than the TagA recording. TagB dropped off sooner—a factor which cannot be controlled. The above discrepancies resulted in shorter time periods before and after the playbacks.

There are a few ways this project could be improved upon, as this was the first time this methodology has been used. The killer whale exposure data may have been more accurate if the tagged pilot whales were not first subjected to sonar. In the future, it would make for better data if the whales were subjected to sonar and killer whale sounds in two separate projects. Also, only one sonar exposure is needed, to prevent the whales from being influenced by one sonar before hearing another. The second sonar exposure was likely at a loudness necessary to truly create an impact on the pilot whales hearing it, as the pilot whales could have confused the first sonar for another whale's call. In addition, data would be more accurate if more replication was achieved by possibly tagging 5-10 pilot whales instead of only two, assuming that the tags stay on for similar amounts of time. This project could also be done with other species of whales, or dolphins, to determine their responses to sonar. It would also help if the movement and activities of tagged whales were tracked so that different behaviors could be detected and the whales' distances from the playback source were known, and whether the whales were sleeping or eating. Also, another factor that should be monitored is whether one of the pilot whales was traveling in a larger group of whales, as this could affect call rates.

So far, there have been few studies on pilot whales and their responses to sonar and other man-made sounds, but other studies on whales have shown that they produce fewer calls when exposed to sonar (Cressey, 2008). Following the hypothesis for this project, it was found that when exposed to sonar, pilot whales generally decrease their average calls/min., but when exposed to killer whale sounds their average calls/minute recovered to near sonar levels. Also, as hypothesized, the average stereotyped call trends closely followed the average call trends, except that during sonar and killer whale sounds the averages decreased to near zero. The pilot whales' responses to killer whale sounds differed from the hypothesis in that during the exposures the average calls/minute increased and then increased again after, but not to the initial averages. The results of this project point towards the conclusion that when whales are exposed to sonar, they produce fewer calls than when exposed to the sounds of natural predators, such as killer whales. As technologies advance and sonar is used more in the ocean, the full adversity of its effects on marine mammals needs to be known.

Acknowledgements

I would like to thank several people who helped me with my experiment. I thank my mother and father, who brought me to WHOI and gave me access to computers and printers for my project. I thank Dr. Peter Madsen for his help providing me with current information on whale decompression sickness. I thank Dr. Peter Tyack for giving me access to the data from his project which I analyzed. I would also like to thank Dr. Muller and Mrs. Centurino for helping me make my graphs.

Research was conducted under NMFS permit #1121-1900 under the authority of the MMPA and the EPA and a permit granted by the Bahamian Government.

Funding for field work was provided by SERDP, ONR, US Navy Environmental Readiness Division (N45) and US Navy Head of Undersea Surveillance (N872A).

Thanks to AUTEC, the NUWC Newport M3R team, Bahamas Marine Mammal Research Organization and members of the BRS team.

Bibliography

American Cetacean Society Authors. 2006. "Pilot Whale".
<http://www.acsonline.org/factpack/PilotWhale.html>

Cardawine, Mark. 2000. Whales, Dolphins and Porpoises. Darling Kindersley, New York, 256 pp.

Cressey, Daniel. 2008. "Sonar Does Affect Whales, Military Report Confirms".
<http://www.nature.com/news/2008/086801/full/news.2008.997.html>

Evans, Peter G.H. 2000. Marine Mammal Biology and Conservation. Platinum Publishers, New York, 630 pp.

Madin, Kate. 2009. Supreme Court Wrestles with Whales vs. Sonar. Oceanus. 47 (2): 42-43

Madsen, Peter T. Professor, University of Aarhus, Aarhus (Denmark). December 6, 2010. Whales and Decompression Sickness. Email to Carlo Bocconcelli (Pers.

Comm.).

Martin, Anthony R. 1990. <u>Whales and Dolphins</u>. Portland House, New York, 192 pp.

McCormack, Claire. 2010. "What Killed the Whales on Ireland's Rutland Island?".
<http://www.time.com/time/health/article/0,8599,2030596,00.html>

Nevala, Amy E. 2009. Investigation Sonar and Whale Strandings. <u>Oceanus</u>. 47 (2): 38-41.

Nicholson, Joseph. 2010. "How Does Sonar Work?".
<http://www.ehow.com/how_does_4567005_Sonar-work.html>

Rossing, Thomas D. 1990. <u>The Science of Sound</u>. Addison-Wesley Publishing   Company, Reading, 686 pp.

Sayigh, Laela. Research Specialist, Woods Hole Oceanographic Institution, Woods   Hole and Teacher at Falmouth Academy, Falmouth. Interview, November 13, 2010.

Tyack, Peter L. 2009. Human Generated Sound and Marine Mammals. Physics Today.   Issue November 2009: 39-44.

Enhancing Algae Cultivation Using an Innovative Photobioreactor

by Aheli Chattopadhyay

Background

In the past two years I worked with reducing carbon dioxide emissions that are believed to contribute to the greenhouse effect.

I merged the topic of CO2 sequestration with another environmental issue, fuel production, which deals with lessening dependency on fossil fuels. I worked with sequestering carbon dioxide to grow photosynthetic algae, which can in turn be used to create fuel.

This year, a homemade photobioreactor (PBR) was built with the incentive of creating a scalable but cost-effective system. The focus was to design a PBR that will be utilized as a test lab set up, but could be scaled up once an optimum design is attained that would yield maximum algae growth.

Purpose

Design a photobioreactor that yields higher algae growth by utilizing a wavelike configuration with optimum fluid agitation such that algae cells are continuously exposed to maximum light, which is provided by a lamp and light reflective mirror.

Also, the photobioreactor should be highly economical in order to be used for lab tests, as well as scalable in order to be used as a commercial PBR.

Hypothesis

The experiment involves three photobioreactors:

1. Traditional horizontal individual tubes
2. Wave-like structure formed from one continuous tube
3. Wave-like structure formed from one continuous tube, used with a light reflective flat mirror

I hypothesize that if all photobioreactors are used for growing algae for the same period of time, then the most algal growth will occur in the third photobioreactor due to greater exposure to sunlight and better fluid mixing.

Materials

- Chemical resistant gloves
- Chemical resistant goggles
- Chlamydomonas reinhardtii algae culture
- Chlorella pyrenoidosa algae culture
- Distilled water
- Airline tubing
- 1.25 in. diameter transparent plastic tubing
- Aquarium air stones
- T-valves
- 2 ft. x 2 ft. flat mirror
- 4 ft. tall 40 watt fluorescent lamps
- Cooking oil dispensers
- Dried Baker's yeast
- Granulated sugar
- 40 watt fluorescent aquarium light
- Graduated cylinder

- Scientific metal measuring spoons
- Spectrophotometer
- F/2 nutrient
- Aquarium air pumps
- Rubber stoppers
- Metal stands
- Wood planks
- Metal clamps
- Check valves
- Cuvettes

Procedure

Photobioreactor 1:

1. Cut three 30 in. tall tubes of 1.25 in. diameter tubing.
2. Orient the two 30 in. long wooden planks at the about 20 in. away from each other.
3. Clamp the three pieces of tubing into place on the planks with 6 supporting metal clamps (3 per plank), each 1 ft. apart.
4. Take two rubber stoppers, and make one hole in one, and two holes in the other.
5. Pass 25 in. airline tubing with attached bubblers through each of the stoppers.
6. Connect the free ends of airline tubing to a T-valve.
7. Attach 20 in. of airline tubing to the free end of the T-valve, and connect the other end of the airline tubing to the aquarium pump.
8. Make a cut about 5 in. from the T-valve in the tubing that connects to the pump, and attach a check valve that points

toward the T-valve.

9. Connect 1 ft. of airline tubing to the tip of the CO2 generator (see How to Make a CO2 generator), and pass 6 of the airline tubing inches through the second hole in the appropriate stopper.

10. Place a fluorescent lamp angled about one foot away from the system.

11. Pour 300 mL water, 30 mL algae culture, and 2 drops f/2 nutrient into a clean quart jar.

12. Pour 110 mL of the quart jar's mixture in through one end of the first tube structure.

13. Pour the other 220 mL of the quart jar's mixture evenly into the other tubes.

14. Cap the tubes with the stoppers.

15. Measure algae growth with a spectrophotometer (see Using a Spectrophotometer).

Photobioreactors 2 & 3:

1. For photobioreactor 2, create a wavelike shape with 8 feet of 1.25 in. diameter transparent plastic tubing with two waves, each with a wavelength of about 1 foot.

2. Orient the two 30 in. long wooden planks at the about 20 in. away from each other.

3. Clamp wave of tubing into place on the planks with 6 supporting metal clamps (3 per plank), each 1 ft. apart.

4. Take two rubber stoppers, and make one hole in one, and two holes in the other.

5. Pass 25 in. airline tubing with attached bubblers through

each of the stoppers.

6. Connect the free ends of airline tubing to a T-valve.

7. Attach 20 in. of airline tubing to the free end of the T-valve, and connect the other end of the airline tubing to the aquarium pump

8. Make a cut about 5 inches from the T-valve in the tubing that connects to the pump, and attach a check valve that points toward the T-valve.

9. Connect 1 ft. of airline tubing to the tip of the CO_2 generator (see How to Make a CO2 generator), and pass 6 of the airline tubing inches through the second hole in the appropriate stopper.

10. Connect 1 ft. of airline tubing to the tip of the CO_2 generator (see How to Make a CO2 generator), and pass 6 of the airline tubing inches through the second hole in the appropriate stopper.

11. Place a fluorescent lamp angled about one foot away from the system.

12. Pour 300 mL water, 30 mL algae culture, and 2 drops f/2 nutrient into a clean quart jar.

13. Pour 165 mL of the quart jar's mixture in through one end of the tube structure

14. Pour the other 165 mL of the quart jar's mixture through the other end of the tubing.

15. Cap the tubes with the stoppers.

16. About six inches away, repeat steps 1-15 for photobioreactor.

17. For photobioreactor 3, place a mirror 6 inches away from

the system, propped 135 degrees above the ground, angled away from the PBR.

18. Measure algae growth with a spectrophotometer (see Using a Spectrophotometer)

How to Make a CO2 Generator

1. In a cooking oil dispenser, use a funnel to pour in 2 tsp of sugar and ½ tsp of yeast.
2. Microwave 250 mL of water under high for 30 seconds.
3. Pour 200 mL of water into the dispenser.
4. Cap the mixture and shake the bottle.
5. Attach of 45 cm of airline tubing to the metal tip.
6. Secure the airline tubing with silicone and let it dry completely before proceeding.
7. At the other end of the tubing, connect a check valve with the arrow pointing in the direction in which the tubing flows into the jar.
8. Connect 55 cm of airline tubing to the airline tubing to the other end of the check valve.
9. Submerge the 55 cm tubing in the algae water.

Method for Analysis: Spectrophotometer

- Beer's Law states that the amount of light absorbed by molecules in a solution is proportional to the amount of the substance doing the absorbance.
- The greater the amount of substance present, the higher the absorbance reading.
- When algae cells in a suspension rather than molecules in a

solution are placed in a spectrophotometer, the absorbance reading actually tells how light is scattered due to the turbidity of the substance. The reading gives a the substance's optical density.

Using a Spectrophotometer

1. To create the test blank, fill a cuvette with distilled water and two drops of F/2 nutrient.
2. Place the test blank in the designated area of the spectrophotometer.
3. To "blank" the instrument, press the zero button.
4. Remove the test blank and place another cuvette in the spectrophotometer, then press the Absorbance button for an absorbance reading.
5. Remove the cuvette and blank the instrument again
6. Repeat steps 1-5 for each algal suspension.

Variables

- Independent variable: PBR Design (Shape and Light Exposure)
- Dependent variable: Algae growth
- Control variable: PBR composed of three horizontal tubes
- Parameter and object constants: Chemical resistant gloves, chemical resistant goggles, amount of algae culture, amount of distilled water, type of plastic tubing, amount of airline tubing, check valves, air stones, T-valves, amount of carbon dioxide, cooking oil dispensers, amount of dried yeast, amount of granulated sugar

Conclusion

It was predicted that over the period of experimentation, the most algal growth would occur in the wave-like PBR formed from one continuous tube placed near a light reflective flat mirror. The hypothesis was correct.

The use of a mirror was effective in increasing the PBR's exposure to sunlight. The wavelike structure was effective in achieving proper fluid mixing. It kept the cells in suspension, helped nutrient distribution, and improved gas exchange. In addition, it eliminated thermal stratification and reduced the risk of photoinhibition, which was necessary because exposure to sunlight increased. The flow rate kept the cells from being placed under high shear stress, which promoted cell growth as opposed to cell damage.

References

Ammann. E.(1967, May). Gas exchange of algae. Retrieved from http /www.ncbi.nlm.nih.gov/pmc/articles/PMC546951/?page=1# pag

Bhamawat, P. (2010, October 20). Growth of chlamydomonas reinhardtii
 under nutrient-limited conditions in steady-state bioreactors. Retrieve
 from http://ecommons.cornell.edu/handle/1813/17689

Chlorella pyrenoidosa. (2008). Retrieved from http://www.ihealthdirectory.com/chlorella-pyrenoidosa/

Csavina, J.L. (2008, August). The optimization of growth rate and lipid content from select algae strains. Retrieved from http://etd.ohiolink.edu/send-pdf.cgi/Csavina%20Janae.pdf?ohiou1215529734

Gebiki. (n.d.). Development of photobioreactors for anoxygeni production of hydrogen by purple bacteria. Retrieved from http://www.aidic.it/pres09/webpapers/163Gebicki.pdf

Greenspan, P. (1985, March). Nile red: a selective fluorescent stain for intracellular lipid droplets.. Retrieved from http://www.ncbi.nlm.nih.gov/pubmed/3972906

Grunwald, M. (2008, April 7). The clean energy scam. Time Magazine

Hirsh, A. (2010, May 19). Diy light reflector. Retrieved from http://www.ehow.com/how_6531743_diy-light-reflector.html#ixzz1DEXdeLkz

Hudock, G. A. (1971), Responses of Chlamydomonas reinhardi to Specific Nutritional Limitation in Continuous Culture. Journal of Eukaryotic Microbiology, 18: 128–131. doi: 10.1111/j.1550-7408.1971.tb03293.x

Hughes, A.P. (1968, August 21). Mutual shading in

quantitative studies
Retrieved from http://aob.oxfordjournals.org/content/33/2/381.short

Massive ice island breaks off greenland. (2010, August 13). Retrieved from http://www.cnn.com/2010/WORLD/americas/08/07/greenland.ice.island/index.html?hpt=T1

Miao, J.R. (2006). A high productivity photobioreactor. Retrieved from http://www.newbioreactor.com/

Molina, E. (2007, February 22). Tubular photobioreactor design for algal cultures. Retrieved from http://www.oilgae.com/blog/2007/02/tubular-photobioreactor-design-for.html

Monastersky, Richard. (1995, July 22). Iron surprise: algae absorb carbon dioxide. Retrieved from http://findarticles.com/p/articles/mi_m1200/is_n4_v148/ai_17352492/

O'Neil, D.L. (n.d.). How to use mirrors to redirect light to plants. Retrieved from http://www.ehow.com/how_5839272_use-mirrors-redirect-light-plants.html#ixzz1GGMu1P1z

Piccolo, T. (2008, December 28). Originoil's bioreactor technology – breakthrough in the production of oil from algae. Retrieved from http://aquaticbiofuel.com/tag/photo-bio-reactor/

Richardson, B. (1969, April 26). Effects of nitrogen limitation on the growth and composition of unicellular algae in continuous culture . Retrieved from http://www.ncbi.nlm.nih.gov/pmc/articles/PMC37795/pdf/applmicro00008-0123.pdf

Tubular photobioreactors. (2001). Retrieved from http://www.oilgae.com/algae/cult/pbr/typ/tub/tub.html

Using the spectrophotometer. (2005). Retrieved from http://webcache.googleusercontent.com/search?q=cache:kKyr3JVtHgJ:faculty.weber.edu/nokazaki/Comparative+Animal+Physiology/Laboratory/Spectrophotometry04-1.pdf+biological+application+of+beer's+law&hl=en&gl=us

Work, V. (2010, August). Increased lipid accumulation in the chlamydomonas reinhardtii sta7-10 starchless isoamylase mutant and increased carbohydrate synthesis in complemented strains. Retrieved from http://ec.asm.org/cgi/reprint/9/8/1251.pdf

Catch that CO^2: Carbon Dioxide Sequestration in Aquifers
By Aheli Chattopadhyay

Abstract
Carbon dioxide is a colorless, odorless green house gas. According to scientists, levels of carbon dioxide in the air are increasing over time and at an accelerating rate. Sequestering this carbon dioxide underground is what many people are turning to. In the process of sequestration, carbon dioxide is caught at the source (power plants) before it is released into the atmosphere. It is then compressed into a supercritical fluid and stored in underground formations or oil and gas fields. One type of underground formations that can sequester carbon dioxide is an aquifer. An aquifer is porous rock filled with water. The layers of an aquifer consist of an unconfined aquifer that opens up to a stream or water source, an impermeable confining layer, and a confined aquifer sandwiched between the confining layer and existing bedrock.

Background
With the threat of global warming, everyone is attempting to find ways to reduce emissions of greenhouse gases such as carbon dioxide. Carbon dioxide sequestration involves storing carbon dioxide underground safely and securely. It is an approach to reducing CO_2 emissions. The focus of this project was to identify the characteristics of an aquifer that

result in the least amount of carbon dioxide leaking out after sequestration.

Problem

What type of confining material (layer of geologic material), is best at preventing CO_2 leakage?

Hypotheses

My two hypotheses are as follows:
- Sand and gravel mixed with clay will form a confining layer that will result in the least carbon dioxide penetration and leakage.
- As the amount of sand and gravel in the confining layer increases, the carbon dioxide leakage decreases. This is because sand and gravel decrease the permeability of the clay, letting less carbon dioxide flow through.

What is CO_2 Sequestration?

- Carbon dioxide sequestration is the long-term storage of supercritical (highly compressed) carbon dioxide in the terrestrial biosphere, underground, or the ocean to slow the build-up of carbon dioxide in the environment.
- Geologic sequestration involves storing carbon dioxide in oil and gas reservoirs, coal seams, or deep saline aquifers.

Aquifers and Confining Layers

- Aquifers are porous and permeable water-bearing rocks located underground. They have enormous storage capacity for carbon dioxide sequestration.
- Confined aquifers are overlain by a confining layer, or unit, which is made up of impermeable rock or clay. The layer limits groundwater movement and prevents carbon dioxide from escaping from an aquifer.

Materials
- Chemical resistant gloves
- Chemical resistant apron
- Chemical splash goggles
- Three 3 liter plastic containers with 3 lids
- Utility knife
- Two 60 cm plastic tubes with an opening of 3.5 cm. in diameter
- 6 ½ liters of distilled water
- 6 lb. of clay
- 1 ½ lb. sand
- ½ lb. gravel
- Timer

Variables
- Independent variables- type of confining material
- Dependent variable- carbon dioxide leakage
- Parameter Constants- time interval, amount of dry ice

- Object Constants- timer, pH probe, amount and type of water, amount and type of clay, type of containers, container lids, safety materials, weight scale, utility knife

Method for Analysis

- The greater the dissolving of CO_2 in water, the lower the pH value of the water, the greater the H^+, and the higher the acidity (since H^+ is an acid). Therefore, a pH meter was used to give pH readings that determined the amount of carbon dioxide that the water absorbed.

Procedure

1. Put on chemical resistant gloves and apron and chemical splash goggles.
2. Mark three containers with 1, 2, and 3 with a marker.
3. Put three containers in a line with six inches between each.
4. Container 1: Use a utility knife to make a hole with a diameter of 3.5 cm. in the middle of the lid.
5. Container 2: Make two large holes, each diameter of 3.5 cm., side by side in the lid (for tubes to go through).
6. Container 3: Make a small hole in the container's lid that is 1 cm. in diameter (for the pH probe). Make another hole in the lid 3.5 cm in diameter (for a tube).
7. Connect containers 1 and 2 by placing a 60 cm. long tube in the holes of the containers' lids.
8. Connect containers 2 and 3 by placing a 60 cm. tube in the holes of the containers' lids.

9. Place slab of confining material in container 2 (refer to "How to make a confining unit").
10. Pour 1 ½ liters of distilled water into container 3.
11. In the small hole of container 3's lid, place the pH probe.
12. Put the lid on container 3.
13. Take a reading of the starting pH of the distilled water.
14. Place ½ liters of dry ice in container 1.
15. Secure all lids.
16. Record pH readings of the water in container 3 every 5 minutes for 45 minutes.
17. Remove the containers from the lids.
18. Clean the tubes connecting the lids with distilled water.
19. Get three new containers, but use the lids from the previous experiment.
20. Re-label the containers and position them the same way.
21. Place a different type of slab of confining material in container 2.
22. Repeat steps 10-16.

How to Make a Confining Layer

Clay-
1. Cut a 3 lb. slab of clay.
2. While putting the clay into container 2, mold it so it stands like a wall in the middle of the container. The wall should be 3.5 cm. thick.

Clay with 25% sand-
1. Cut a 2.25 lb. slab of clay.
2. Measure out 0.75 lb. of pure sand.

3. Measure ½ liter of water.
4. Mix and knead the clay with sand, using water to get them to stick together.
5. When you put the clay into container 2, mold it so it stands like a wall in the middle of the container. The wall should be 3.5 cm. thick.

Clay with 50% sand:
1. Follow the procedure for clay with 25% sand, except using 1.5 lb. of clay and 1.5 lb. sand.

Clay with 25% sand & gravel:
1. Follow the procedure for clay with 25% sand, except using 2.25 lb. clay, 0.375 lb. sand, and 0.375 lb. gravel.

Clay with 50% sand & gravel:
1. Follow the procedure for clay with 25% sand, except using 1.5 lb. clay, 0.75 lb. sand, and 0.75 lb. gravel.

Results
- My first hypothesis was that sand and gravel mixed with clay will form the confining layer with the least CO_2 leakage.
- My results do support this hypothesis. According to my results, the layer made of the clay with sand and gravel was best at preventing CO_2 leakage.
- My second hypothesis was that as the amount of sand and gravel in the confining layer increases, the carbon dioxide leakage decreases.

- My results support this hypothesis as well. The 50% sand and gravel confining layer allowed the least amount of CO_2 leakage.

Data Analysis

- The following are the confining layers in order from which allowed the most carbon dioxide leakage to which allowed the least leakage.

   1. Clay alone
   2. Clay/25% sand
   3. Clay/50% sand
   4. Clay/25% sand & gravel
   5. Clay/50% sand & gravel

- When clay alone was tested, the water came out to be the most acidic, meaning the most carbon dioxide was absorbed by the water. This means more carbon dioxide passed through the confining layer due to a higher permeability in the clay wall.

- When only sand was added to the clay, less carbon dioxide was able to penetrate the confining layer.

- When the sand and gravel confining layers were tested, the water was the least acidic, and the least carbon dioxide was absorbed by the water. Less carbon dioxide passed through the confining layer due to a lower permeability in the wall.

- The 50% clay with sand and gravel gave the best results of all the confining layers tested, so the sequestration would be most effective in an aquifer with a confining layer is one made up of much sand and gravel.

Conclusions
- Clay's porosity is 50-70%. However, it has very low permeability because the pores are very small and not interconnected well enough to admit fluids to pass through. Gravel and sand have 20-50% porosity, yet they are more permeable than clay because the pores are interconnected much more.

- In this experiment, porosity is not a huge concern. The amount of pores is not what affects the carbon dioxide penetration; it's how well those pores are connected to let fluids pass through.

According to the above, it seems that since clay is less permeable than sand and gravel, clay alone would be the best confining unit. However, my results show that clay with sand and gravel would form the best unit.

The following is my reasoning behind my results.
- The small clay particles fill up the pore spaces between the sand and gravel so that the pores in the sand and gravel are no longer as connected.

- The small size, flatness, and lack of connectivity of its particles makes clay have a low permeability. When adding more small particles to the clay (sand and gravel), it becomes even more difficult to penetrate.

- It seems that when you combine the clay, sand, and gravel, the clay works to balance the sand and gravel out. With the materials working together in this way, the overall permeability of the confining unit is even lower than clay alone as the confining unit.

If confining unit has low permeability, it means the carbon dioxide will have a hard time passing through the unit and leaking out.

The project was based on what type of confining material would let the least amount of carbon dioxide leak through. An experiment was set up to demonstrate the confining layer of an aquifer with carbon dioxide trying to penetrate it. On one side of the confining unit, carbon dioxide was released, and on the other side of the confining unit, the carbon dioxide was captured in water; the amount that was able to penetrate the unit was then measured. The less carbon dioxide that passed through the confining layer, the less the pH of the water changed. The data showed that the sand with gravel and clay aquifer was the best at sequestering carbon dioxide. Thus, carbon dioxide sequestration will have the most lasting effects when the gas is sequestered into a sand-and-gravel aquifer with a confining layer made up of clay, sand, and gravel.

References

Ammann. E.(1967, May). Gas exchange of algae. Retrieved from http://www.ncbi.nlm.nih.gov/pmc/articles/PMC546951/?page=1#pag

Bhamawat, P. (2010, October 20). Growth of chlamydomonas reinhardtii under nutrient-limited conditions in steady-state bioreactors. Retrieved from http://ecommons.cornell.edu/handle/1813/17689

Monastersky, Richard. (1995, July 22). Iron surprise: algae absorb carbon dioxide. Retrieved from http://findarticles.com/p/articles/mi_m1200/is_n4_v148/ai_17352492/

Richardson, B. (1969, April). Effects of nitrogen limitation on the growth and composition of unicellular algae in continuous culture . Retrieved from http://www.ncbi.nlm.nih.gov/pmc/articles/PMC377951/pdf/applmicro00008-0123.pdf

Work, V. (2010, August). Increased lipid accumulation in the chlamydomonas reinhardtii sta7-10 starchless isoamylase mutant and increased carbohydrate synthesis in complemented strains. Retrieved from http://ec.asm.org/cgi/reprint/9/8/1251.pdf

Emissions into Energy: Algae Biofuel Through Biosequestration
2009-2010 - Year 2
By Aheli Chattopadhyay

Abstract 2009-2010

Levels of carbon dioxide in the air have been increasing, contributing to the greenhouse effect. To prevent global warming, researchers have developed a method called sequestration. This involves storing the carbon dioxide in secure areas, but the safest storage source is unknown. Another global concern is that of fuel production. The world depends on non-renewable resources for energy. Because fast-growing algae produce many lipids that can be extracted and used for energy, algal biofuel is currently a great topic of interest. In an interest to merge these two environmental issues, a project was designed based on how different amounts of carbon dioxide affect algal growth.

The Experiment

The experiment involved the creation of a homemade photobioreactor, an artificial photosynthetic environment. The original algae culture was placed in glass quart jars, with a fluorescent bulb hanging overhead the jars. The control had an airstone bubbling passive air through the jar's saltwater mixture. A carbon dioxide generator was made using yeast, water, and sugar in a oil dispenser. The carbon dioxide produced by alcoholic fermentation was piped into each jar of the experimental group.   Different amounts of yeast and sugar were used in order to produce varying amounts of carbon dioxide. The data showed that more carbon dioxide increased algal growth. However, after a certain point, carbon dioxide was unable to continue increasing the growth rate of algae. In an actual setting, after this point, sequestering CO_2 to make algae in a photobioreactor would be no longer cost-effective. By establishing this extreme, a relationship was developed between the carbon dioxide absorbed and the resulting algae growth; it was predicted that 34 liters of CO_2 could produce 35.35 kg of algae.

Purpose
Determine the amount of carbon dioxide that affects algae growth in a homemade photobioreactor, and establish a relationship between the two factors.

Hypothesis
According to the photosynthesis equation, carbon dioxide and water are used by algae to produce oxygen and glucose. While the algae grows and undergoes photosynthesis, I hypothesize that increased carbon dioxide will accelerate growth.

Materials
- Chemical resistant gloves
- Chemical resistant goggles
- 40 mL Tetraselmis Chui algae culture

- 1 gallon of distilled water
- 2 quart jars
- Ocean salt
- Airline tubing
- 2 aquarium check valves
- 2 aquarium air stones
- T-valve
- Cooking oil dispenser
- Dried Baker's yeast
- Granulated Sugar
- Mass scale
- 40 watt fluorescent aquarium light
- Graduated cylinder
- Scientific metal measuring spoons
- Silicone sealant

Procedure

1. Put on safety goggles and chemical resistant gloves and apron.
2. Place 235mL of distilled water in each of the two quart jars.
3. Add 1.5 tsp of ocean salt to the water.
4. Place the lids on the jars and shake the saltwater mixture to fully dissolve the salt.
5. Connect 45 cm of airline tubing to the aquarium air pump.
6. To the other end of the tubing, attach a check valve with the arrow pointing in the direction in which the tubing flows into the jar.

7. Attach 40 cm of airline tubing to the other end of the check valve.
8. Link a T-valve to the 40 cm tubing.
9. To other two barbs of the T-valve, attach two pieces of 25 cm airline tubing.
10. Connect airstones to the end of each piece.
11. Submerge one airstone in jar 1's saltwater solution, leave the second airstone outside of jar 2, then plug in the pump.
12. Pour 20 mL of Tetraselmis Chui algae culture into each jar.
13. Immediately add 1mL of f/2 nutrient to each jar.
14. Mass the jars with their contents on a scale and record.
15. Place an overhead florescent light above the aquarium, keeping it on 18 hours during the day and off 6 hours at night. During the 6 hours when the light is off, place the second airstone in jar 2.
16. Leave jar 1 as is.
17. Place the free end of the CO_2 generator's airline tubing in jar 2.
18. Mass the jars after two weeks.
19. Repeat the experiment for multiple trials with different ratios of yeast and sugar.

Variables
- Independent variable: Amount of CO_2 (mL)
- Dependent variable: Algae growth (g)
- Control variable: Algae grown by passive air exchange (setup one)

•Parameter and object constants: Chemical resistant gloves, chemical resistant goggles, amount of Tetraselmis Chui algae culture, amount of distilled water, quart jars, amount of ocean salt, amount of airline tubing, check valves, air stones, T-valve, amount of light, cooking oil dispenser, amount of dried yeast, amount of granulated sugar, silicone sealant

Production of CO2 in a CO2 Generator
After extensive research, I have established the following relationship of CO_2 production from yeast.

My sources	My experiment
40g granulated sugar	2 tsp granulated sugar
10g dry active yeast	0.5 tsp yeast
544ml of CO2	6.8 mL of CO2

- 1 gram = 4 tsp
- 40g sugar = 160 tsp sugar
- 10g yeast = 40 tsp yeast

I used 2 tsp of sugar and 0.5 tsp yeast, or 1/80 of the average amount of materials my sources used.

Data Analysis
The following are the average amounts of algae growth from least to greatest and the average carbon dioxide absorption.
1. 2.36g algae –passive air

2. 3.07g algae –3.40mL CO_2
3. 5.80g algae –5.10mL CO_2
4. 7.07g algae –8.50mL CO_2
5. 6.77g algae –7.65mL CO_2
6. 6.10g algae –6.80mL CO_2

The control produced the least carbon dioxide. Thus as the amount of CO_2 increases, algae yield increases.

- The experiment with 6.8mL of CO_2 exhibited maximum algae growth. With further increase in carbon dioxide, algae production stays within the same range. This means that after a certain point, additional carbon dioxide does not increase algal growth rate as a result of photosynthesis-related factors. Because injecting carbon dioxide into the photobioreacter and growing the algae itself is an expensive process, this information is important in determining how much carbon dioxide to use to produce maximum amount of algae.
- With my data, I can establish that in a larger photobioreactor, 35.35 kg of algae can be produced by absorbing 34 liters of CO_2.

Conclusions

- I hypothesized that algae growth would accelerate with additional carbon dioxide. The results support my hypothesis.
- I established a relationship between the CO_2 absorbed by the algae during photosynthesis and amount of algal growth.
- I developed the initial criteria needed to design an effective, economical photobioreactor that will grow algae for biofuel production.

References

Biofuel guide-ethanol. (2008, February 17). Retrieved November 3, 2009, from http://biofuelguide.net/

Bouck, J. (2009, January 1). *The Basic reactor, algae geek.* Retrieved November 3, 2009, from http://algaegeek.com/aboutus.aspx

Chinongoza, M. (2007, November 12). *How to Grow algae fo Biodiesel.* Retrieved October 14, 2009, from http://ezinearticles.com/?How-To-Grow-Algae-For-Biodiesel&id=829645

Cohen, J, & Hopwood, N. (2000). *Greenhouse gases and society.* Retrieved November 3, 2009, from http://www.umich.edu/~gs265/society/greenhouse.htm

Evision Holdings (2005). *Why Algae biodiesel?.* Retrieved November 3, 2009, from http://www.whyinvestinenergy.com/why-algae-biodiesel.php

Ewers, J, & Wiechers, G. (2009, March 1). Algae: putting carbon dioxide in a bind. *Power Engineering International, 17*(3), Retrieved November 12, from http://www.powergenworldwide.com/index/display/articledisplay/356755/articles/power-engineering-international/volume-17/issue-3/germany-focus/algae-putting-carbon-dioxide-in-a-bind.html

FAQ Information Portal (2009, May 14). *Netl: are the environmental or safetry concerns related to carbon sequestration?.* Retrieved December 10, 2009, from http://www.netl.doe.gov/technologies/carbon_seq/FAQs/concerns.ht

Growing algae. (2010). Retrieved November 22, 2009, from http://www.growing-algae.com/algae-media.html

Heath, D, & Trigg, L. (2008). *Do-it-yourself carbon dioxide injection.* Retrieved October 21, 2009, from http://www.aquabotanic.com/DIYCO2.htm

Jander, G. (2009, June 13). *CO2 in the aquarium.* Retrieved November 3, 2009, from http://fins.actwin.com/mirror/plant-co2.html

Liebert, T. (2008, April/May). *CO2 sequestration by algae reactors.* Retrieved October 3, 2009, from http://kansas.sierraclub.org/Planet /2008-0405/AlgaeCO2Sequestration.pdf

Lenntech. (2009). *Algae description and types.* Retrieved Octobe 21, 2009, from http://www.lenntech.com/eutrophication-water-bodies/algae.htm

Levine, J., & Miller, K.R. (2004). *Biology.* Upper Saddle River, NJ: Pearson Prentice Hall.

Miller, R.S. (2001, June 10). *An Algae experiment.* Retrieved November 24, 2009 from http://fins.actwin.com/aquaticplants/month.200106/msg00185.html

Sachs, P. (2010). *The Start of the Food Chain- microalgae culturing.* Retrieved September 14, 2009, from http://www.aquaculturestore.com/info/microal.html

Schirber, M. (2009). Powerful ideas: wringing oil from algae.

LiveScience, Retrieved September 12, 2009, from http://www.livescience.com/environment/090406-pf-algae-fuel.html

Shepherd, Roy. (2010). *Discovering fossils*. Retrieved September 30, 2009, from http://www.discoveringfossils.co.uk/fossilfuels.htm

Oilgae. (2008). *Transesterification*. Retrieved November 3, 2009, from http://www.oilgae.com/algae/oil/biod/tra/tra.html

Investigating the Effects of Greywater on Landscape Agriculture
By Gregory Corbett

Introduction

Water is one of Earth's most precious natural resources and it is necessary for the survival of most living organisms. Many human beings have misconceptions about the availability of water and have a complete disregard for its conservation, resulting in its misuse and overuse. Today, with easy access to clean water, the average American uses between 302.83 liters (80 gallons) and 378.54 liters (100 gallons) of water a day which results in approximately 98,420,706,384 liters (26 billion gallons) of water being used daily in the United States (Flowers 1). In some parts of the United States, as much as half of the water used is for outdoor watering (McLamb 6). Today, the problems of water shortages due to drought conditions, increased population, and human activities are occurring all over the world. Reusing wastewater produced in homes is one way to help meet the ever increasing water needs. Graywater, or wastewater generated by the sink, shower, tub, washing machine, and dishwasher, could potentially substitute freshwater in many aspects of our lives. Studies have shown that in many areas of the United States, approximately one-third of a household's water consumption is used on lawns (Weber 1). Investigating the use of graywater irrigation on landscape agriculture, namely turf grasses, will provide valuable data as to the effect this reused water has on the typical lawns and whether graywater is a neglected resource worth tapping into.

Purpose

"…the planet is in the midst of what the United Nations is calling a water crisis" (Clark 1). Drought conditions and the ever increasing population have put high demands on our freshwater supply. Finding ways to conserve and reuse water, one of Earth's most precious resources, is vital. Studies have shown that as much as half of the water consumed by household's in the United States is used for outdoor watering (McLamb 6). The purpose of this experiment is to investigate the effect that graywater, household wastewater from the shower, dishwasher, and washing machine, has on one type of landscape agriculture, turf grasses.

Hypothesis

Based on the research conducted prior to the experiment, it was learned that graywater, namely household wastewater generated from the shower/tub, dishwasher, and washing machine, is more than ninety-nine percent water and less than one percent organic particles which are typically microscopic in size. This organic material consists of detergent, salts, body oils, hair and skin follicles, dirt, and micro-fibers (McLamb 7). The typical topsoil environment contains oxygen, bacteria, nematodes, and earthworms which help to breakdown this type of organic material into microscopic elements that can be absorbed by and beneficial to plants (McLamb 8). Therefore, it is hypothesized that the graywater from the shower, dishwasher, and washing machine will not negatively affect the root or blade growth of the turf grasses when compared to those watered with the tap water control. In fact, it is hypothesized that the microscopic elements, nutrients, in the graywater will be beneficial to the turf grasses and they will show improved growth over the turf grasses irrigated with the tap water control.

Materials

- 16 – Plastic planting trays 33 cm x 38 cm x 10 cm
- 1- galvanized steel professional soil sampler (takes a 2.54 cm core sample)
- Portable soil pH meter from Carolina Biological Co.
- 4- 250 ml plastic graduated cylinders
- Spectral Strip pH strips, pH 0-14, 100 pack
- LaMotte NPK (nitrogen, phosphorus, potassium) soil test kit
- 2- sod samples bluegrass 152.4 cm x 61 cm Tuckahoe Farms of New England
- 2-sod samples Versaturf (fine fescue) 152.4 cm x 61cm Tuckahoe Farms
- 4 – 2 L plastic containers
- 4- cool white fluorescent lights 750 lumens each and 67 cm long
- 4- 18 kg bags Premium Top Soil

- 2- 20 kg bags of crushed white stone
- 1 pair of industrial shears for cutting sod and blades
- Disposable gloves
- 16 liters distilled water
- White plastic tablecloths to cover tables
- VWR 300g electronic balance
- 1- 30 cm ruler

Procedures

1) **Procedure for preparing sod**

The 33cm x 38 cm x 10 cm growing trays were rinsed with distilled water. Then the trays were filled half way with Premium Top Soil. The top soil was firmly pressed into the trays. Four trays were prepared for the bluegrass and 4 trays were prepared for the Versaturf. The sod was then cut with professional shears and firmly fitted and pressed into the top soil in each tray. Each tray was then placed into another tray which was 1/3 filled with crushed stone for drainage. After the initial data was obtained, the samples were irrigated with 500 mL water once a week. One tray for each sod type was irrigated with shower water, one tray for each sod type with washing machine water, and one tray with dishwasher water for each sod type. The fourth tray for each sod type was the control and was watered with tap water.

2) **Procedure for measuring blade height**

Using a 30 cm plastic ruler, the average blade height was measured. The height was measured from the top rim of the growing tray to the top of the blades of the turf grass. The height was not measured from the top of the soil since soil compresses with watering and would not give a consistent measurement.

3) Procedure for obtaining soil and turf core sample

Holding the soil sampling tube in the vertical position, it was forced into the sod through the soil. The tube was twisted back and forth to sever the core. The tube with the core was then carefully lifted from the planting tray. The core sample was removed from the sampling tube. The plant growth (blade growth) was cut using professional shears from the root growth at the soil line. The fresh blade weight was obtained using a VWR 300 g electronic balance. The root growth was washed under running tap water for approximately 20 seconds to remove the loose soil. The sample was patted dry and the fresh root growth was weighed using the electronic balance. The blade and root growth was then allowed to dry for 24 hours at room temperature. The dry blade and root weight was then obtained.

4) Procedure for obtaining soil pH

When the core sample was removed with the soil sampler, the portable pH meter probe was injected into the soil of the core sample and the pH was measured and recorded.

5) Procedure for obtaining the nitrogen, phosphorus, and potassium levels in the soil

These measurements were obtained using the Lamotte Soil NPK Test Kit. The extraction tube from the test kit was filled with 30 ml of distilled water. Two Floc-Ex tablets provided in the kit were dissolved in the distilled water. A sample of soil was then added to the solution and shaken for one minute. The tube was allowed to sit and settle for 5 minutes.

To test for nitrogen -10 ml of the clear solution above the soil was added to another vial and one Nitrate test tablet was added. The solution was shaken and after 5 minutes, the pink color that developed was compared to the nitrogen color chart provided.

To test for phosphorus - 25 drops of the clear solution above the soil was added to another vial and the vial was filled to the 10 ml mark with distilled water. One Phosphorus tablet was added and the solution was shaken. After 5 minutes the blue color that developed was compared to the phosphorus color chart provided.

To test for potassium - 10 ml of the clear solution above the soil was added to another vial and one Potassium tablet was added. The solution was shaken and after 5 minutes the cloudiness of solution was compared to the potassium color chart provided.

6) Procedure for obtaining the pH of the water samples

The pH of the water samples was obtained using the Spectral Strip pH strips. The strips were dipped into the water sample for 10 seconds. The strips were carefully shaken to remove the excess water and the color change was compared to the color chart provided.

Calculations

Since roots allow a plant to absorb water and nutrients from the soil, a healthy root system is key to a healthy plant. The root to shoot ratio is one measure to help assess the overall health of a plant. The turf grass irrigated with the tap water is the control which will give a "normal' root to shoot ratio. Any changes, either up or down, from this normal level would be an indication in the change of the health of the turf grass.

Calculation for Determining the Root to Shoot Ratio

Weigh and record the root and blade weights (dry) for each sample. The calculation for the root to shoot ratio is:

Dry weight of root / Dry weight of top of plant (blade)

Week 0 (11/28/2010) Initial Data

Sod Sample = Bluegrass
Water type: Ratio:
Dishwasher - 0.8g/0.5g =1.60
Wash Machine – 0.9g/0.5g=1.80
Shower –0.8g/0.5g = 1.60
Tap – 1.0g/ 0.6g = 1.66

Sod Sample = Versaturf
Water type: Ratio:
Dishwasher - 0.9g/0.3g = 3.00
Wash Machine – 0.7g/0.2g = 3.5
Shower –1.0g/0.3g = 3.33
Tap –0.9g/0.3g = 3.00

Week 1 (12/5/10) Data

Sod Sample = Bluegrass
Water type: Ratio:
Dishwasher: 1.0g/0.9g = 1.11

Sod Sample = Versaturf
Water type: Ratio:
Dishwasher: 1.0g/0.3g = 3.33

Washing Machine: 1.3g/0.8g =1.63
Shower: 1.1g/0.7g = 1.57
Tap: 1.2g/ 0.7g =1.71

Washing Machine: 0.7g/0.2g = 3.50
Shower: 1.0g/0.3g = 3.33
Tap: 0.9g/0.3g = 3.00

Week 2 (12/12/10) Data

Sod Sample = Bluegrass
Water type: Ratio:
Dishwasher: 1.0g/1.0g = 1.00
Washing Machine: 1.6g/1.0g = 1.60
Shower: 1.3g/0.8g = 1.63
Tap: 1.4g/0.8g = 1.75

Sod Sample = Versaturf
Water type: Ratio:
Dishwasher: 0.9g/0.3g =3.00
Washing Machine: 0.9g/0.3g =3.00
Shower: 1.2g/0.4g = 3.00
Tap: 1.3g/0.4 =3.25

For Week 3 (12/19/2010) Data

Sod Sample = Bluegrass
Water type: Ratio:
Dishwasher: 1.0g/1.1g = 0.91
Washing Machine: 1.9g/1.1g = 1.72
Shower: 1.5g/0.9g = 1.66
Tap: 1.5g/0.9g = 1.66

Sod Sample = Versaturf
Water type: Ratio:
Dishwasher: 0.9g/0.3g = 3.00
Washing Machine: 1.5g/0.4g =3.75
Shower: 1.5g/0.5g =3.00
Tap: 1.5g/0.5g =3.00

Week 4 (12/26/2010) Data

Sod Sample = Bluegrasss
Water type: Ratio:
Dishwasher: 1.0g/ 1.0g= 1.00
Washing Machine: 2.3g/1.3g= 1.76
1.9g/0.5g = 3.80
Shower: 1.9g/1.1g = 1.72

Sod Sample = Versaturf
Water type: Ratio:
Dishwasher: 0.9g/0.3g = 3.00
Washing Machine:
Shower: 1.9g/0.6g = 3.16

Tap: 1.9g/1.1g = 1.72   **Tap**: 1.9g/ 0.6g =3.16

Week 5 (1/02/2011) Data

Sod Sample = Bluegrass   **Sod Sample** = Versaturf
Water type:Ratio: **Water type**: Ratio:
Dishwasher: 1.1g/1.1g = 1.00   **Dishwasher**: 1.0g/0.4g = 2.50
Washing Machine: 2.5g/1.4 = 1.78   **Washing Machine**: 2.3g/0.6g =3.83
Shower: 2.2g/1.3g = 1.69   **Shower:** 2.1g/0.6g = 3.50
Tap: 2.3g/1.3g = 1.77   **Tap**: 2.2g/0.6g = 3.66

For Week 6 (1/09/2011) Data

Sod Sample= Bluegrass   **Sod Sample** = Versaturf
Water type: Ratio: **Water type: Ratio:**
Dishwasher: 1.0g/1.2g =0.83   **Dishwasher:** 1.1g/0.5g =2.20
Washing Machine: 2.7g/1.5g = 1.80 **Washing Machine**: 2.5g/0.6g = 4.16
Shower: 2.4g/1.4g = 1.71   **Shower**: 2.4g/0.7g =3.42
Tap: 2.5g/1.4g = 1.78   **Tap**: 2.5g/0.7g =3.57

For Week 7 (1/16/2011) Data

Sod Sample: Bluegrass   Sod Sample: Versaturf
Water type: Ratio   **Water type: Ratio**
Dishwasher: 1.1g/1.3g = 0.84   **Dishwasher**: 1.2g/0.6g =2.00
Washing Machine: 3.0g/1.6g = 1.88 **Washing Machine**: 2.8g/0.7g =4.00
Shower: 2.7g/1.5g = 1.80   **Shower**: 2.6g/0.7g = 3.71
Tap: 2.8g/1.6g = 1.75   **Tap**: 2.6g/0.7g = 3.71

Observations

Within 24 hours of planting the sod in the appropriate planting trays and watering with the graywater or tap water, all the turf grasses appeared slightly wilted and possibly in shock. After 72 hours of receiving approximately 12 hours of grow light per day, the turf grasses appeared healthier. The blades were no longer wilted and were standing up straight. The soil was still moist in all samples.

On December 5, 2010, 7 days after the initial planting, the turf grasses all appeared healthy and an increase in blade growth height was observed. It was also observed on this date that the Lamotte N, P, K test procedure was tedious and long. The color change that occurred in each vial was very subjective. The soil in all samples was just starting to look dry and all were watered with appropriate water after the testing was complete.

Over the next 7 days, blade growth in terms of height was clearly visible on all samples. It was observed that the bluegrass contained a broader blade and it was greener and had a shiner tone than the Versaturf. The Versaturf was a much finer blade of grass and had a much lighter green coloring. The Versaturf samples do not appear to have as healthy of growth in terms of shoot growth as the bluegrass.

On December 12, 2010, further core samples were taken. The Versaturf samples did not appear as healthy in terms of blade growth as the bluegrass samples. Fine white new root growth was clearly visible on the Versaturf shower, tap, and washing machine samples and also on all the bluegrass samples. Over the next 7 days, the blade growth in all the bluegrass samples was visibly increasing. The blades of this turf grass were noticeable broader and greener than the Versaturf. The Versaturf samples were showing blade growth, however, since the blades were so fine, they did not appear to be growing as well as the bluegrass.

On December 19, 2010, the N, P, K, testing showed an increase in phosphorus in the soil of both the bluegrass and Versaturf. It was observed that the bluegrass sample watered with the washing machine water showed slightly stronger blade height growth than those watered with the tap or shower water. The blade growth of the bluegrass watered with the dishwasher water appeared not as strong in growth in terms of height as the other bluegrass samples. This observation was confirmed by the data and calculations of the root to shoot ratios. The Versaturf samples all appeared approximately equal in terms of blade growth.

During the week leading to December 26, 2010, the bluegrass samples continued to show strong blade growth. The height of all the blades was increasing. The bluegrass sample watered with washing machine water appeared to have slightly higher blade growth, however, the tap and shower samples appeared very close in height. The bluegrass and the Versaturf samples watered with dishwashing water appeared to have the least blade growth and overall health. On December 26, 2010, it was observed that the phosphorus levels in the samples watered with the washing machine water increased to 15ppm.

Over the next few weeks, the bluegrass samples watered with the tap, shower, and washing machine water all continued to show increased blade height growth. The blades also began bending toward the grow lights. The blades growth was increasing so that the blades were also leaning over. The dishwasher sample appeared to have the least growth. The Versaturf samples did not appear to have the same strong blade growth in term of height or color as the bluegrass. The phosphorus levels increased slightly in the washing machine soil samples and also began to show an increase in the shower soil samples. Strong new root growth continued to be seen in both the bluegrass and Versaturf samples watered with the shower, tap, and washing machine water. Some new root growth, but not as much as in the other specimens, was seen the dishwasher specimens.

Date	Water Source	Fresh Weight: Blade (g)	Root (g)	Blade (g)	Root (g)	Shoot Ratio

Date	Fixture					
28-Nov	Tap	1.5	3.0	0.6	1.0	1.66
28-Nov	Dishwasher	2.8	2.8	0.5	0.8	1.60
28-Nov	Washing Machine	2.5	3.0	0.5	0.9	1.80
28-Nov	Shower	2.4	3.0	0.5	0.8	1.60
5-Dec	Tap	1..6	3.1	0.7	1.2	1.71
5-Dec	Dishwasher	3.0	3.0	0.9	1.0	1.11
5-Dec	Washing Machine	2.4	3.1	0.8	1.3	1.63
5-Dec	Shower	2.4	3.1	0.7	1.1	1.57
12-Dec	Tap	1.8	3.2	0.8	1.4	1.75
12-Dec	Dishwasher	3.0	2.8	1.0	1.0	1.00
12-Dec	Washing Machine	2.4	3.3	1.0	1.6	1.60
12-Dec	Shower	2.6	3.1	0.8	1.3	1.63
19-Dec	Tap	2.2	3.2	0.9	1.5	1.66
19-Dec	Dishwasher	3.1	2.8	1.1	1.0	0.91
19-Dec	Washing Machine	2.9	3.4	1.1	1.9	1.72
19-Dec	Shower	2.8	3.2	0.9	1.5	1.66
26-Dec	Tap	2.6	3.3	1.1	1.9	1.72
26-Dec	Dishwasher	3.1	2.9	1.0	1.0	1.00
26-Dec	Washing Machine	3.1	3.5	1.3	2.3	1.76

Date						
26-Dec	Shower	2.92	3.3	1.1	1.9	1.72
2-Jan	Tap	2.83	3.4	1.3	2.3	1.77
2-Jan	Dishwasher	.1	3.0	1.1	1.1	1.00
2-Jan	Washing Machine	3.33	3.6	1.4	2.5	1.79
9-Jan	Shower	3.0	3.3	1.3	2.2	1.69
9-Jan	Tap	2.93	3.5	1.4	2.5	1.78
9-Jan	Dishwasher	.2	3.0	1.2	1.0	0.83
9-Jan	Washing Machine	3.63	3.7	1.5	2.7	1.80
9-Jan	Shower	2.3	3.4	1.4	2.4	1.71
16-Jan	Tap	3.0	3.6	1.6	2.8	1.75
16-Jan	Dishwasher	3.	3.1	1.3	1.1	0.84

| 16-Jan | Washing Machine | 23.6 | 3.7 | 1.6 | 3.0 | 1.88 |

Date	Tap	Dishwasher	Washing Machine	Shower
28-Nov	9.0	9.0	8.0	8.0
5-Dec	10.0	9.0	10.0	9.0
12-Dec	14.0	11.0	15.0	13.0
19-Dec	17.0	12.0	18.0	16.0
26-Dec	22.0	14.0	23.0	21.0
2-Jan	23.0	16.0	25.0	22.0
9-Jan	25.0	17.0	27.0	24.0
16-Jan	27.0	19.0	31.0	28.0

Blade Height: Bluegrass

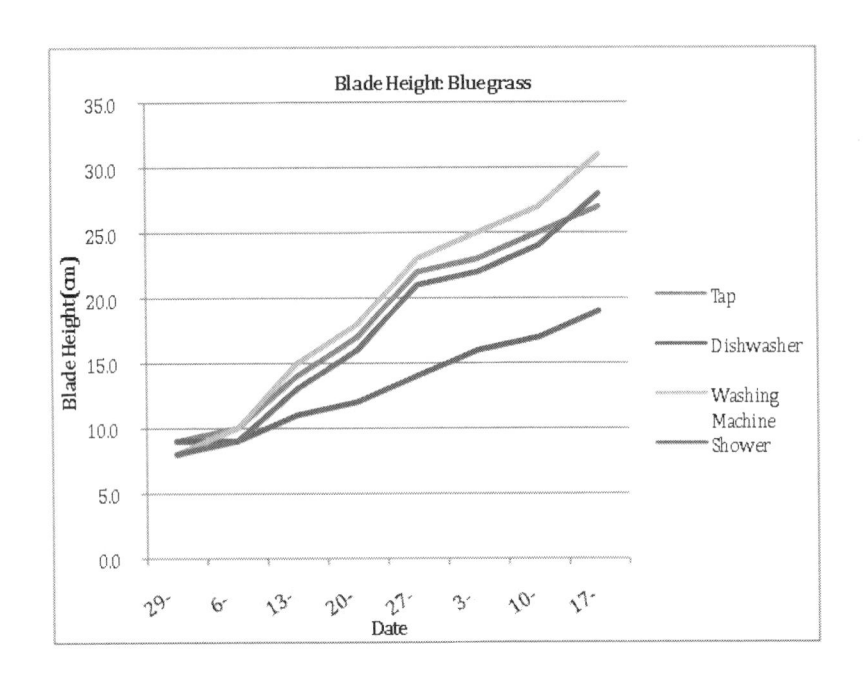

Versaturf Data						
Date	Water Source	Fresh Weight: Blade (g)	Fresh Weight: Root (g)	Dry Weight: Blade (g)	Dry Weight: Root (g)	Root to Shoot Ratio
28-Nov	Tap	1.3	2.5	0.3	0.9	3.00
28-Nov	Dishwasher	1.4	2.5	0.3	0.9	3.00
28-Nov	Washing Machine	0.9	2.2	0.2	0.7	3.50
28-Nov	Shower	1.3	3.0	0.3	1.0	3.33
5-Dec	Tap	1.6	2.6	0.3	0.9	3.00
5-	Dishwasher	1.3	2.6	0.3	1.0	3.33

Date	Appliance					
Dec 5-Dec	Washing Machine	0.8	2.6	0.2	0.7	3.50
Dec 5-Dec	Shower	1.3	3.2	0.3	1.0	3.33
Dec 12-Dec	Tap	1.7	2.6	0.4	1.3	3.25
Dec 12-Dec	Dishwasher	1.3	2.5	0.3	0.9	3.00
Dec 12-Dec	Washing Machine	0.8	2.6	0.3	0.9	3.00
Dec 12-Dec	Shower	1.8	3.2	0.4	1.2	3.00
Dec 19-Dec	Tap	1.8	2.9	0.5	1.5	3.00
Dec 19-Dec	Dishwasher	1.3	2.5	0.3	0.9	3.00
Dec 19-Dec	Washing Machine	1.4	3.0	0.4	1.5	3.75
Dec 19-Dec	Shower	2.1	3.5	0.5	1.5	3.00
Dec 26-Dec	Tap	2.0	3.0	0.6	1.9	3.17
Dec 26-Dec	Dishwasher	1.5	2.5	0.3	0.9	3.00
Dec 26-Dec	Washing Machine	1.7	3.0	0.5	1.9	3.80
Dec 26-Dec	Shower	2.3	2.6	0.6	1.9	3.17
Dec 2-Jan	Tap	2.2	3.1	0.6	2.2	3.67
2-Jan	Dishwasher	1.6	2.6	0.4	1.0	2.50

Date	Source					
2-Jan	Washing Machine	2.0	3.4	0.6	2.3	3.83
2-Jan	Shower	2.3	3.6	0.6	2.1	3.50
9-Jan	Tap	2.4	3.2	0.7	2.5	3.57
9-Jan	Dishwasher	1.7	2.7	0.5	1.1	2.20
9-Jan	Washing Machine	2.2	3.6	0.6	2.5	4.17
9-Jan	Shower	2.4	3.7	0.7	2.4	3.43
16-Jan	Tap	2.4	3.4	0.7	2.6	3.71
16-Jan	Dishwasher	1.8	2.8	0.6	1.2	2.00
16-Jan	Washing Machine	2.4	3.6	0.7	2.8	4.00
16-Jan	Shower	2.5	3.9	0.7	2.6	3.71

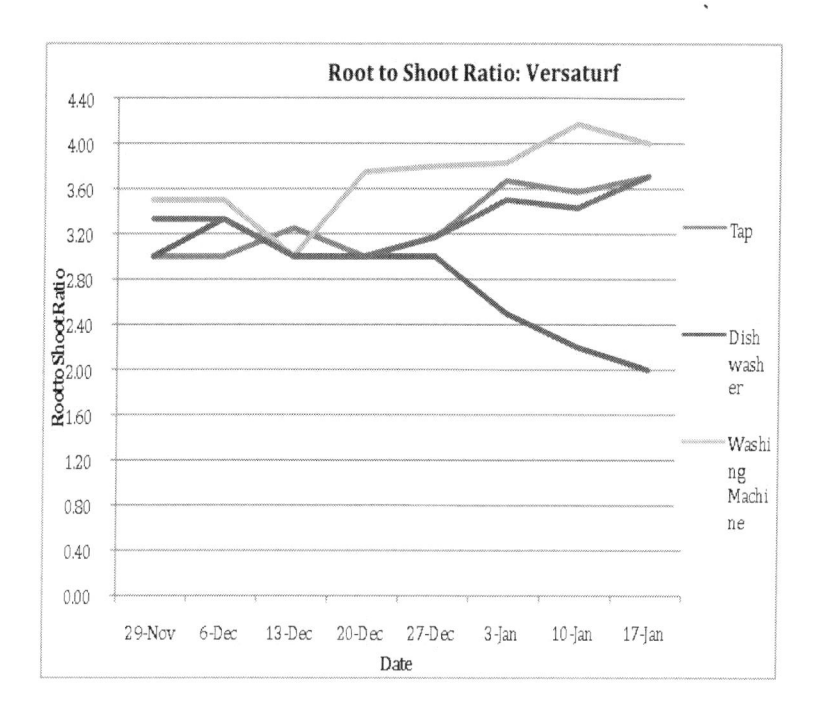

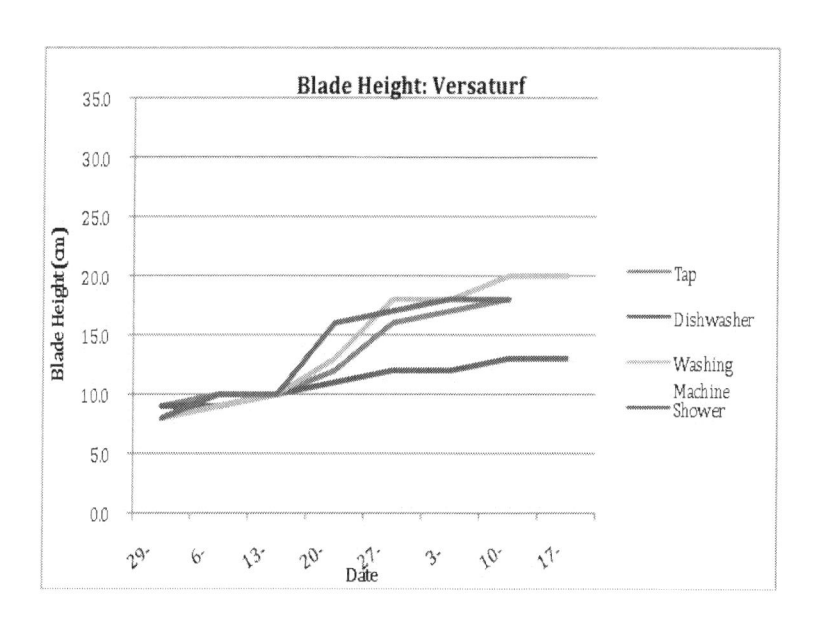

Blade Height (in cm): Versaturf

Date	Tap	Dishwasher	Washing Machine	Shower
28-Nov	9.0	9.0	8.0	8.0
5-Dec	10.0	9.0	9.0	10.0
12-Dec	10.0	10.0	10.0	10.0
19-Dec	12.0	11.0	13.0	12.0
26-Dec	16.0	12.0	18.0	16.0
2-Jan	17.0	12.0	18.0	17.0
9-Jan	18.0	13.0	20.0	18.0
16-Jan	18.0	13.0	20.0	18.0

Soil pH: Blue Grass

Date	Tap	Dishwasher	Washing Machine	Shower
28-Nov	7.0	7.0	7.0	6.8

Date				
5-Dec	7.1	7.0	7.0	6.9
12-Dec	7.0	7.0	7.0	6.8
19-Dec	7.1	7.4	7.1	6.7
26-Dec	7.1	7.6	7.0	6.8
2-Jan	7.0	8.0	7.0	6.7
9-Jan	7.1	8.5	7.0	6.7
16-Jan	7.0	8.7	7.0	6.7

Soil pH: Bluegrass

Soil pH: Versaturf

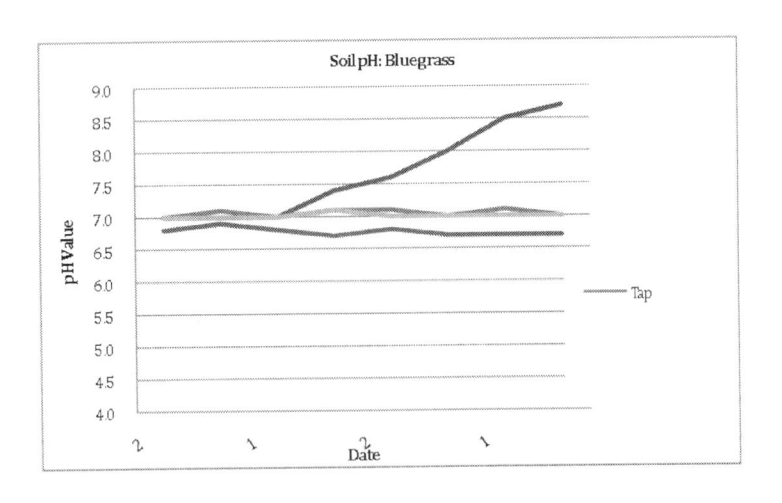

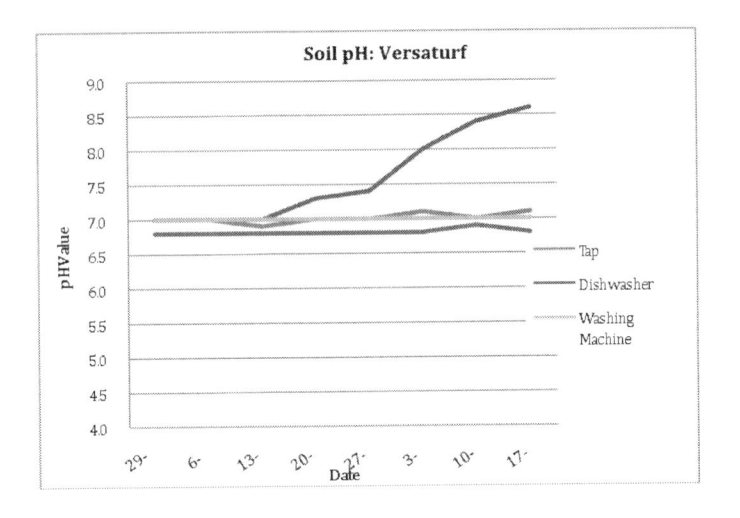

Soil pH: Versaturf

Date	Tap	Dishwasher	Washing Machine	Shower

Date				
28-Nov	7.0	7.0	7.0	6.8
5-Dec	7.0	7.0	7.0	6.8
12-Dec	6.9	7.0	7.0	6.8
19-Dec	7.0	7.3	7.0	6.8
26-Dec	7.0	7.4	7.0	6.8
2-Jan	7.1	8.0	7.0	6.8
9-Jan	7.0	8.4	7.0	6.9
16-Jan	7.1	8.6	7.0	6.8

Date	Tap	Shower
28-Nov	7.5	6.5
5-Dec	7.5	7.0
12-Dec	7.0	6.5

Date		
19-Dec	7.5	6.5
26-Dec	7.5	7.0
2-Jan	7.0	6.5
9-Jan	7.5	7.0
16-Jan	7.0	6.5

Date	Water Source	Nitrogen	Phosphorus	Potassium
28-Nov	Tap	20	4	40
28-Nov	Dishwasher	20	4	40
28-Nov	Washing Machine	20	4	40
28-Nov	Shower	20	4	40
5-Dec	Tap	20	4	40
5-Dec	Dishwasher	20	4	40
5-Dec	Washing Machine	20	4	40
5-Dec	Shower	20	4	40
12-Dec	Tap	20	4	40
12-Dec	Dishwasher	20	4	40
12-Dec	Washing Machine	20	4	40
12-Dec	Shower	20	4	40
19-Dec	Tap	20	4	40
19-Dec	Dishwasher	20	4	40
19-	Washing	20	12	40

Date	Appliance			
19-Dec	Machine			
26-Dec	Shower	20	4	40
26-Dec	Tap	20	4	40
26-Dec	Dishwasher	20	4	40
26-Dec	Washing Machine	20	15	40
2-Jan	Shower	20	4	40
2-Jan	Tap	20	4	40
2-Jan	Dishwasher	20	4	40
2-Jan	Washing Machine	20	15	40
9-Jan	Shower	20	4	40
9-Jan	Tap	20	4	40
9-Jan	Dishwasher	20	4	40
9-Jan	Washing Machine	20	15	40
16-Jan	Shower	20	12	40
16-Jan	Tap	20	4	40
16-Jan	Dishwasher	20	4	40

Jan				
16-Jan	Washing Machine	20	20	40
16-Jan	Shower	20	12	40

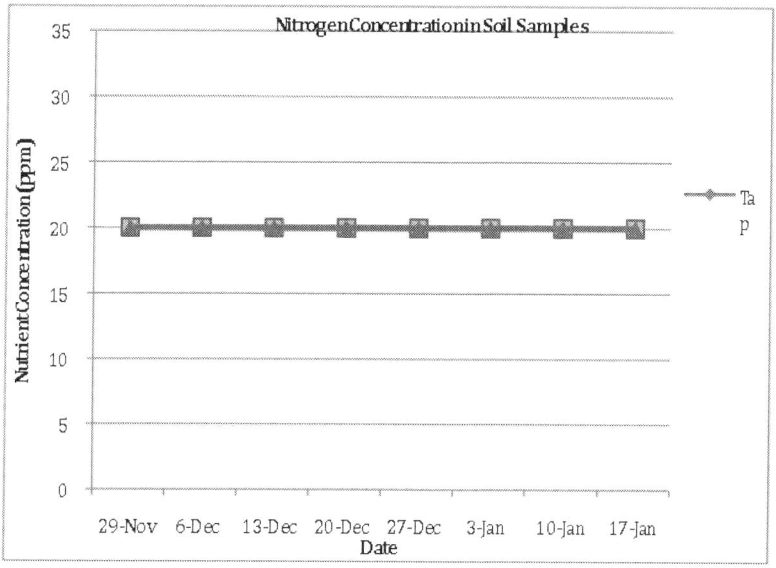

NitrogenConcentrationinSoil Samples
Nutrient Concentration (ppm)
35
30
25
20
15
10
5
0
29-Nov
6-Dec
13-Dec
20-Dec
27-Dec
3-Jan
10-Jan
17-Jan
Date
Tap

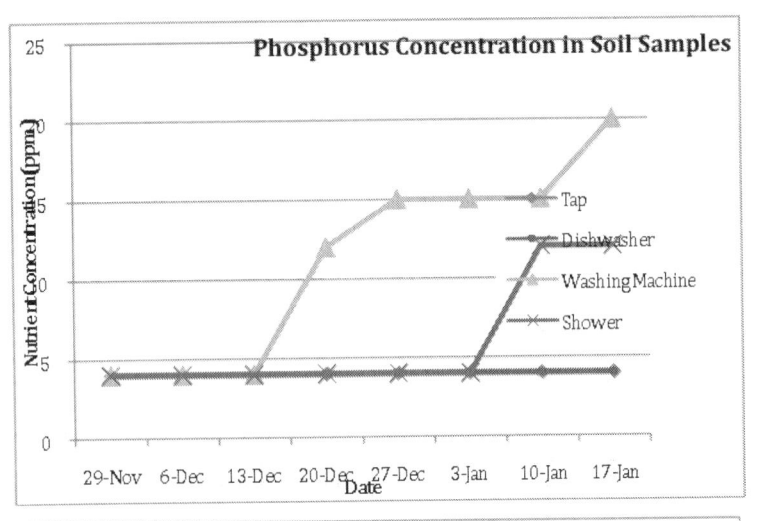

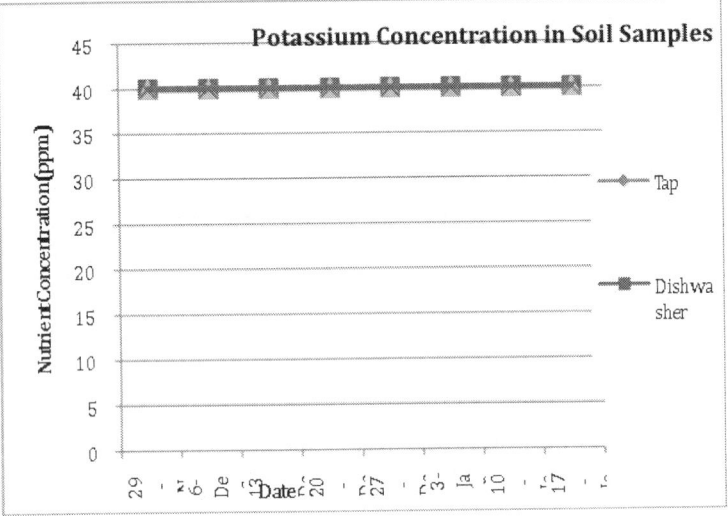

Analysis of Data

The data was collected at T_0, the initial planting date of the sod samples and then once a week (every 7 days) for 7 weeks. Core samples were analyzed each week for blade and root weight, blade height, water pH, soil pH, and soil nutrient levels.

The bluegrass and the Versaturf (fine fescue) species of turf grasses irrigated with tap water (control), shower water, and washing machine water all exhibited an increase in blade height, blade weight, root weight, and, thus, root to shoot ratios over the testing period. The blue grass sample irrigated with tap water had a blade weight increase from 0.6g to 1.6g (167%), a root weight increase from 1.0g to 2.8g (180%), a blade height increase from 9cm to 27cm (200%), and a root to shoot ratio increase from 1.66 to 1.75 (5.42%). The blue grass sample irrigated with washing machine water had blade weight increase from 0.5g to 1.6g (220%), a root weight increase from 0.9g to 3.0g (233%), a blade height increase from 8cm to 31cm (288%), and a root to shoot ratio increase from 1.80 to 1.88 (4.44 %). The blue grass sample irrigated with the shower water had a blade weight increase from 0.5g to 1.5g (200%), a root weight increase from 0.8g to 2.7g (238%), a blade height increase from 8cm to 28cm (250%), and a root to shoot ratio increase from 1.60 to 1.80 (12.5%). The blue grass sample irrigated with the dishwasher water had a blade weight increase from 0.5g to 1.3g (160%), a root weight increase from 0.8g to 1.1g (38%), a blade height increase from 9cm to 19cm (111%), and a root to shoot ratio decrease from 1.60 to 0.84 (-90%).

The Versaturf sample irrigated with tap water had a blade weight increase from 0.3g to 0.7g (133%), a root weight increase from 0.9g to 2.6g (189%), a blade height increase from 9cm to 18cm (100%), and a root to shoot ratio increase from 3.00 to 3.71 (23.7%). The Versaturf irrigated with the washing machine water had a blade weight increase from 0.2g to 0.7g (250%), a root weight increase from 0.7g to 2.8g (300%), a blade height increase from 8cm to 20cm (150%), and a root to shoot ratio increase from 3.50 to 4.00(14.3%). The Versaturf irrigated with the shower water had a blade weight increase from 0.3g to 0.7g (133%), a root weight increase from 1.0g to 2.6g (160%), a blade height increase from 8cm to 18cm (125%), and a root to shoot ratio increase from 3.33 to 3.71(11.4%). The Versaturf irrigated with the dishwasher water had a blade weight increase from 0.3g to 0.6g (100%), a root weight increase from 0.9g to 1.2g (33.3%), a blade height increase from 9cm to 13cm (44.4%), and a root to shoot ratio decrease from 3.0 to 2.0 (-50%). The pH of the soil irrigated with the tap water, shower water, and washing machine water all remained in the 6.8 to 7.5 range, which is ideal for turf grass. However, pH of the dishwasher water was approx. 11.0 which caused the pH of the soil irrigated with the dishwasher water to increase from 7.0 to 8.7 over the course of the experiment. The nitrogen soil levels all remained at approximately 20ppm over the course of the experiment. The potassium soil levels also continuously remained at approximately 40ppm. The phosphorus levels remained at approximately 4ppm for the soil irrigated with the tap and dishwasher water. However, the soil irrigated with the washing machine water exhibited an increase in the phosphorus level from 4ppm to 20ppm over the course of the experiment. The soil irrigated with the shower water also showed an increase in the phosphorus level from 4ppm to 12ppm.

Water analysis for an outside laboratory, BWA Water Additives, confirmed the pH of the water results. Also, phosphate analysis was performed which indicated that the phosphate level in the tap water was 0.02ppm, the dishwasher water was 0.06ppm, the shower water was 0.1ppm, and the washing machine water was 0.2ppm. This increase in the phosphate levels corresponds with the higher blade and root growth weights and blade heights seen in the sod samples watered with washing machine and shower water. The independent water analysis also tested sulfate levels and conductivity and found a very large increase in ion levels for the dishwater water. Since sulfates and salinity in general are damaging to soil and plant life, and disrupt the roots ability to absorb nutrients, this data combined with the high pH could explain the slow growth rate and decrease in the root to shoot ratio compared to the control of the sod irrigated with the dishwasher water.

Conclusions

Based on the data obtained, the hypothesis made for this experiment was partially correct. The bluegrass and Versaturf (fine fescue) species of cool weather turf grass both demonstrated growth that was similar to the control group (tap) water when irrigated with graywater from the shower. These species also demonstrated slightly improved growth patterns compared to the control group (tap) when irrigated with graywater from the washing machine. Growth patterns were measured in terms of dry blade weight, dry root weight, blade height, and root to shoot ratios. It was correctly hypothesized that the increase in nutrient level, phosphorus, in the washing machine graywater likely led to the improved growth pattern. Although it was hypothesized that dishwasher graywater would also have a positive impact on turf grass growth, this was not the case. The dishwasher graywater negatively impacted the growth of both species most likely due to the high pH, high sulfate, and high conductivity (ion) levels which research has shown leads to an increase in salinity levels of the soil. Increased soil salinity is detrimental to root and plant growth since it is acutely toxic and interferes with the root's ability to absorb water and nutrients.

Relevance

Water shortages, droughts, increased populations, graywater, and beautiful lawns? Can these items really go together? This research and experimentation proved they can. Water is one of Earth's most precious and vital natural resources and finding ways to conserve and reuse it are crucial. Reusing the wastewater or graywater produced in a household from the sink, shower/tub, and washing machine is one way to meet the ever increasing water needs.

This research showed the turf grasses, namely bluegrass and fine fescue, exhibited no adverse effects in terms of growth when irrigated with graywater generated from the shower/tub. In fact, these species exhibited slightly improved growth patterns compared to the control group (tap water) when irrigated with graywater from the washing machine. This research also demonstrated that the graywater generated from the dishwashing water negatively impacted the turf grass growth, which correlates with the literature findings that dishwasher water should not be considered a graywater but rather a blackwater.

Research has shown that the average United States household (family of three) consumes 454, 250 liters (120,000 gallons) of water a year. It is estimated that an average family generates between 189, 271 liters (50,000 gallons) – 227,125 liters (60,000 gallons) of graywater a year. Studies have shown that one-third of a household's water usage, approximately 151,416 liters (40,000 gallons), is for irrigation of outside lawns. Therefore, the average family generates more than enough graywater to irrigate their lawns thus greatly cut their demand for fresh water. Graywater produces a "new" water supply since it allows water that was previously wasted to be beneficially used. This greatly reduces a household's fresh water intake and also then reduces the amount of wastewater sent to sewer or septic systems (McLamb 6).

Works Cited
Brzuszek, Robert F. "Home Landscape in Mississippi." *Mississippi State University*. Web. 6 Jan. 2011. <http://msucares.com////.html>.

BWA Water Additives. *Lab Analysis for Dishwasher Water*. Wood Dale. Print.

- - -. *Lab Analysis for Shower Water*. Wood Dale. Print.

- - -. *Lab Analysis for Tap Water*. Wood Dale. Print.

- - -. *Lab Analysis for Washing Machine Water*. Wood Dale. Print.

Clark, Josh. "Exactly What Happens If We Run Out of Water." *How Stuff Works*. 2010. Web. 13 Oct. 2010. <http://science.howstuffworks.com///geophyisics/out-of-water.htm>.

Flowers, Bryan. "Domestic Water Conservation: Greywater, Rainwater, and Other Innovations." *Proquest*. 2004. Web. 23 Oct. 2010. <http://www.cas.com>.

LSU AgCenter. *Dealing With Salinity*. Web. 18 Jan. 2011. <http://text.lsuagcenter.com>.

McLamb, Curtis. *Graywater: The Next Wave. New Solutions for the Old Problems of Water Shortage and Septic System Failure*. Suwanee: Graywater Resource INC., 2004. Print.

New Jersey Agricultural Experiment Station. "Managing Soil pH for Turfgrasses." *Rutgers Cooperative Research and Extension*. Web. 5 Dec. 2010. <http://www.rce.rutgers.edu>.

Procter and Gamble. "Tide Liquid." *Tide*. 2010. Web. 7 Jan.

2011. <http://www.tide.com>.

- - -. "Tide Powder Detergent." *Tide*. 2010. Web. 6 Jan. 2011. <http://www.tide.com>.

Savewater. "Gardening with Greywater-Soil and Plant Information." *Savewater*. 2005. Web. 21 Jan. 2011. <http://www.savewater.com.au>.

- - -. "Greywater Friendly Plants." *Savewater*. 2005. Web. 21 Jan. 2011. <http://www.savewater.com.au>.

- - -. "Managing Salinity When Irrigating with Greywater." *Savewater*. 2005. Web. 21 Jan. 2011. <http://www.savewater.com.au>.

Science Buddies. "Measuring Plant Growth." *Science Buddies*. 2011. Web. 5 Dec. 2010. <http://www.scienceuddies.org>.

Tuckahoe Farms INC. "Tuckahoe Turf Farms." *Tuckahoe Farms INC.* N.p., 2007. Web. 28 Oct. 2010. <http://www.ttfarms.com>.

Tuckahoe Turf of New England. "Tuckahoe Turf Farms Installation." *Tuckahoe Turf of New England*. 2009. Web. 5 Dec. 2010. <http://www.tuckahoeturf.com>.

- - -. "Tuckahoe Turf Products." *Tuckahoe Turf of New England*. 2009. Web. 28 Nov. 2010. <http://www.tuckahoeturf.com>.

U.S. Department of Health and Human Services. "Household Products Database-Health and Safety Information on Household Products." *U.S. Department of Health and Human Services*. Web. 7 Jan. 2011.
<http://householdproducts.nlm.nih.gov>.

Weber, Christopher. "Going Gray: Trim Your Water Footprint-and Bill- with Gray Water." *Green Living: House and Home*. Web. 23 Oct. 2010.
<http://www.emagazine.com>.

Weblife. "Chapter 9: Alternative Grey Water Systems." *Humanure Handbook*. 1-11. Web. 23 Oct. 2010.
<http://weblife.org/humanure/.html>.

Tooth Whitening
By Kari Cristoforo

"The problem that I was trying to solve is, 'What brand of toothpaste will make your teeth the whitest in the shortest amount of time (12 day period of time), Ultra Brite, Toms of Maine, or Colgate Total?' In order to investigate this question I set up an experiment that would use eight tiles for each group to see what brand made the largest difference in the starting/ending tile color. While doing this experiment I also used paint examples that I compared each color to day by day."

Literature Review

Before I conducted my experiment, I had to figure out what products to use in order to stain the non-glossed tiles. As I researched further and further, I found out that the four worst products for your teeth when it comes to stains are black tea, coffee, cola, and red wine. Black tea is a terrible teeth stainer, and some tea alternatives include green, white and herbal teas which are actually good for you. Coffee has quite a discoloring effect on your teeth, due to it being caffeine rich it'd be hard to give up but some ways to minimize the discoloring effect include drinking it through a straw or diluting it with milk and brushing/chewing gum after drinking it.

When it comes to cola, it's something that everyone loves and drinks all the time. Little does everyone know that cola drinks not only stain and discolor your teeth but also erode enamel and cause tooth decay due to the harmful phosphoric acid, citric acid and sugar content in them. You can minimize some of these side effects by simply drinking it with a straw. Red wine contains polyphenols that stain your teeth and the alcohol in red wine is very acidic so it wears the enamel of the teeth just as cola does. After you consume red wine, **do not** brush your teeth, but rather rinse your mouth and chew gum.

Along with choosing what products I want to stain the tiles with, I also had to choose what products I was going to whiten the teeth with. After researching what products of whitening toothpaste work best, I came up with Colgate Total (whitening), Ultra Brite Advanced Whitening, and Toms of Maine Natural Whitening Toothpaste. Not only would testing these three products tell me what one works best, but it will also inform me of whether using organic toothpaste will actually make a difference.

Experts recommend Colgate Total more often than any other toothpaste, and this was the first toothpaste to include the antibacterial ingredient triclosan. Triclosan fights gum disease, and remains active between brushings. Ultra Brite is budget toothpaste that actually beat out 40 other toothpastes for stain removal. It contains fluoride and a tartar-control ingredient but doesn't contain triclosan that makes cavity protection last between meals. Toms of Maine is much less abrasive than the precious two toothpastes but it is said to not remove stains as well, and is rather costly due to being organic.

The reason that I choose to use a non-glossed tile rather than a tooth is mainly because I couldn't get teeth. So, I researched what may be a close alternative and I found out that non-glossed tiles will actually absorb stains and other chemicals, which is what my experiment involved. The reason that it will absorb stains is because they are generally white (so the stains become noticeable), and because there isn't a gloss or anything covering the stone, so it's basically just white stone, which would definitely absorb anything!

Tooth whitening is a procedure that lightens teeth and helps to remove stains and discoloration. Whitening is among the most popular cosmetic dental procedures because it can significantly improve the appearance of your teeth at much less cost and inconvenience than other techniques. Did you know that every day a thin coating forms on your teeth and it picks up stains? Also, the outer layer of each tooth (enamel) contains pores that hold stains!

Your teeth can become discolored by stains on the surface or by changes in the tooth material. There are three types of discoloration, extrinsic discoloration, intrinsic discoloration, and age-related discoloration. Extrinsic discoloration is when the outer layer of the tooth (the enamel) is stained by coffee, wine, cola, or other drinks/foods. Intrinsic discoloration is when the inner structure of the tooth (the dentin) darkens or gets a yellow tint. Causes include excessive exposure to fluoride during early childhood. This type of discoloration usually may be more difficult and take longer to remove the stains. Age-related discoloration is a combination of extrinsic and intrinsic factors. In addition to stains caused by foods/smoking, the dentin naturally yellows over time. The enamel that covers the teeth gets thinner with age, which allows the dentin to show through. People generally get stains on their tooth enamel by using tobacco, drinking coffee, tea, colas, or red wine, ingesting too much fluoride when teeth are forming, and even having had trauma to your teeth. Some symptoms include stains on the enamel or a yellow tint in the dentin. There is no special diagnosis; a dentist can simply diagnose tooth discoloration by looking at the teeth.

Some tooth discoloration can be removed with professional cleaning, but many stains are permanent unless the teeth are treated (whitened) with a bleaching agent. Some may wonder how you prevent all these processes from occurring. Well, it's as simple as brushing your teeth after every meal, rinsing your mouth with water after having food/drinks that stain your teeth, and setting-up regular cleanings with your dentist. If you are unhappy with the appearance of your teeth, call your dentist so they can reassure that you get the whitening treatment that's right for you!

Almost anyone whose permanent teeth are in can qualify for tooth whitening. Ask your dentist which whitening technique is best for you, because the procedures vary based on where your stains are (enamel or dentin), and how dark they are. Teeth that are yellow respond best to bleaching. Brown or gray teeth may not whiten evenly when bleached. People with periodontal disease (or sensitive teeth), may want to avoid chemical whitening techniques that can irritate tender gums.

Over a decade of research has proven bleaching and other whitening methods to be both safe and effective. Sensitivity may occur in people after whitening procedures, particularly when they eat hot or cold foods, but usually disappears after 48 hours and stops completely when treatment is stopped. Although, in a few cases your dentist may discourage dental bleaching, such as if you have gum disease, teeth with worn enamel, cavities or particularly sensitive teeth, if you're pregnant or breast feeding, or if you have crowns, caps, or other dental work in your mouth that can't be bleached. Whitening procedures may take place in the dental office or at your own house.

There are two main types of whitening procedures. When whitening is done on a tooth that has had root-canal treatment and no longer has a live nerve, the process is called non-vital whitening. Vital whitening means that the procedure is being done on teeth that have live nerves. During non-vital whitening procedures, he or she will place a whitening agent inside the tooth and will place a temporary filling. It will be left this way for several days. You may need this done only once, or it can be repeated until the tooth reaches the desired shade. During a vital whitening procedure, a gel-like whitening solution (which usually contains hydrogen peroxides) is placed in a tray that resembles a night guard or mouth guard. The tray is then placed over the teeth for a certain period of time-anywhere from an hour or two to overnight. Your dentist will also apply a special gel to the gums to protect them from the whitening agent (this is done before the whitening agent is applied).

All whitening techniques work in one of two ways, a bleaching procedure or a non-bleaching procedure. Bleaching procedures change your natural tooth color, usually anywhere from five to seven shades brighter. Bleaches contain an active ingredient, most often carbamide peroxide or hydrogen peroxide in concentrations of 10-22%, which helps remove both deep and surface stains. Whitening toothpastes have special chemical or polishing agents that provide additional stain removal. But, remember that everyone responds differently to different whitening procedures!

Enamel
Dentin
Gum
Pulp
Periodontal ligament
Bone
Crown
Root

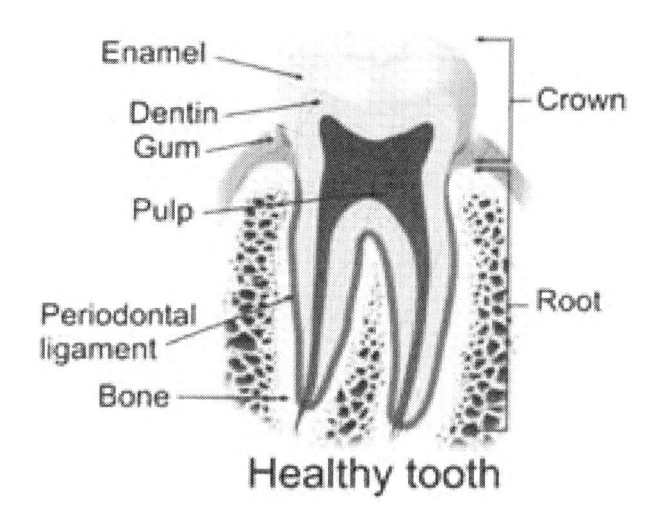

Healthy tooth

Each tooth is composed of multiple layers and components. The enamel is the hardest substance in the human body. Underneath the enamel lies the dentin, a layer that is hard but not quite as hard as the overlaying enamel. Dentin extends down into the roots, and within the dentin are millions of tiny tubes that run into the next layer, called the pulp. The pulp is the central portion of the tooth, and consists of blood vessels and nerves. It is the layer that nourishes the tooth, and despite its appearance, teeth are living structures. A bone-like substance called cementum covers the roots, and finally each tooth is held in position and anchored to bone by a periodontal ligament. The basics of tooth decay is the bacteria that inhabit the mouth. A great number of species of bacteria live in the mouth, where they convert foods- especially sugars and starches-into acids. The bacteria combine with food debris, saliva and other compounds in the mouth to form a coating on the teeth. The coating is referred to as plaque, and is normally removed by brushing and flossing your teeth. Within tartar the acids from the metabolic activity of the bacteria can dissolve the enamel, which is another reason whitening toothpaste is beneficial-because it strengthens the enamel of the teeth.

It is important to have good enamel and take care of your enamel because it can't be repaired or replaced once it is severely harmed. Tooth enamel is made almost entirely of rod-shaped crystals of calcium phosphate, packed tightly together. This is the heaviest substance in the body and doesn't contain cells nor need a blood supply. Compared to enamel, dentin has a more open structure, but it is still heavier and harder than most types of bone (not harder/heavier than enamel). Dentin is produced by cells in the pulp cavity, and it needs a blood supply to stay alive.

(On the right is the enamel, and on the left is the dentin). Remember that teenagers shouldn't have their teeth bleached until they are between 14 and 16 years old. This is because in a younger child, the nerve of the tooth (pulp) hasn't fully developed, and whitening this could irritate this/cause sensitivity. Also, bleaching doesn't leave teeth permanently white, although it can last from 6 months to a year.

The Effects of Tree bark Extracts on Jurkat T and HeLa Cells

By Kayleigh Fay, Bruce Bausk, and Taylor Fearing
Mentor: Prof. Thomas C. Chiles

Introduction

Cancer is an illness affecting many people all over the world. Recent studies have found that extracts from trees could possibly be used to treat patients with this life-threatening disease. This experiment was inspired by the aforementioned studies and aims to further investigate possible uses of tree extracts as cancer treatments. The topics pertinent to this experiment include cells, cancer, trees, useful chemicals found in trees, equipment used, extraction, and Cisplatin, among others. The extracts that will be used are from the bark of *Annona muricata* (graviola), *Cinnamomum verum* (cinnamon), *Betula papyrifera* (paper birch), *Prunus Serotina* (black cherry), and *Tabebuia avellanedae.* The two types of human tumor cells that will be investigated are the following: HeLa cells and Jurkat T cells.

Problem Statement

The purpose of this experiment is to determine the effectiveness of tree bark used in conjunction with and in place of the chemotherapy drug Cisplatin (*cis*-diamminedichloroplatinum (II)) to inhibit growth or cause death of cancer cells.

Research
Cells

The cell is the fundamental unit of life; all living organisms are composed of cells. There are two types of cells: prokaryotic and eukaryotic. Prokaryotic cells have no nucleus, while eukaryotic cells have a nucleus. The basic components of a eukaryotic cell are the cell membrane, nucleus, and organelles. The membrane is a phospholipid bilayer that is selectively permeable, meaning it allows only certain types and sizes of molecules to enter and exit the cell. The nucleus controls all functions that occur in the cell. It also contains the cell's DNA (deoxyribonucleic acid), or genetic information. The organelles present in a cell vary depending on the function of the cell. An organelle is a small, membrane-enclosed structure designed to perform a specific task within the cell. Examples of organelles include mitochondria, which convert glucose to ATP to provide cells with energy, and peroxisomes, which break down toxic peroxides. Cells of complex organisms undergo specialization, meaning they develop specifically for the task they will perform, and therefore have varying types and proportions of organelles (Sadava, Hillis, Heller, & Berenmaum, 2011).

Cell Communication

There are three main steps to cell communication: reception, transduction, and response. In reception, a signal molecule binds to a receptor protein, causing it to change configuration. This binding is highly specific. Most receptors are proteins; some are intracellular, found in the nucleus or cytosol of large target cells. Other receptors include activated hormone-receptor complexes, which can act as a transcription factor, turning on and off specific genes. Three main receptors are G-protein-linked receptors, receptor tyrosine kinases, and ion channel receptors. G-protein-linked receptors are plasma membrane receptors that work with the help of a G-protein (protein with a specific "G" structure), which acts as an on/off switch. Receptor tyrosine kinases are membrane receptors that attach phosphates to tyrosines, an amino acid used to synthesize proteins. An ion channel receptor acts as a gate when a signal bonds to it, allowing specific ions, such as Na+ through a channel in the receptor (Campbell, Reece, & Urry, 2005).

Transduction is generally a multi-step process that involves cascades of molecular interactions which relay signals from receptors to target molecules in the cell. The molecules that first relay a signal are generally proteins known as first messengers.

These proteins act as dominoes, causing a chain reaction within the cell.

At each step there is a conformational change caused by phosphorylation, the adding of a phosphate group, which activates second messengers.

Second messengers are small, non-protein, water-soluble molecules or ions. The most widely used second messengers are cyclic AMP (cAMP), Calcium ions and inositol triphosphate (IP3). Cyclic AMP stands for cyclic adenosine monophosphate, and is converted from the energy molecule ATP, or adenosine triphosphate, by a membranous enzyme. When acting as a second messenger, cAMP activates a protein kinase which phosphorylates (or adds a phosphate group) and then triggers other cellular responses. Calcium ions are effective messengers because cells can regulate their concentrations. When first messengers stimulates a G protein linked receptor, the G protein activates the enzyme phospholipase C which cleaves a membrane protein into IP3. The inotisol triphosphate is translocated to the endoplasmic reticulum (an organelle of the endomembrane system in a cell) and is attached to an IP3 gated ion channel, releasing Ca2+ ions which activate other cellular responses (Karp, 1999).

Responses to signals lead to the regulation of cellular activities such as cell division, cell death, and the activation or inhibition of enzymes.

The Cell Cycle

The eukaryotic cell cycle of somatic, or body, cells consists of two phases: mitosis and interphase. The movement of the cell between these two phases is controlled by a sort of "cellular clock" which varies for each cell (Smith & Martin, 1973). The "clock" has specific checkpoints where the cell cycle stops until it receives a "go-ahead" signal. At these checkpoints, the cell must be large enough, have correctly replicated DNA, have all chromosomes (genetic material), and be correctly attached to microtubules in preparation for division. All external growth factors (such as having adequate space to divide) must also be achieved for the cell to continue in the cell cycle (Campbell, Reece, & Urry, 2005).

Cells spend the about 90% of the cycle in interphase; during interphase the cell is not dividing. Interphase is divided into three parts. G1 phase, also called gap one phase, is the first growth phase and the site of the first checkpoint. During this phase, the cell grows and replicates its organelles. G1 seems to be the most important checkpoint; if the cell does not pass the checkpoint, it will go into a Go phase, in which it is neither growing nor dividing (Smith & Martin, 1973). If the cell passes the G1 checkpoint, it then proceeds to the S (synthesis) phase. Here, DNA is replicated before moving on to G2, or gap two, phase. In the G2 phase, the cell continues to grow and replicate its organelles (Campbell, Reece, & Urry, 2005). Upon completing G2, the cell will move into Mitosis. The purpose of mitosis is to divide nuclear material. Mitosis is facilitated by a centromere, a complex of microtubules that controls the movement of the nuclear materials during the process by attachment to kinetochores on the center of the chromosomes (Campbell, Reece, & Urry, 2005).

Mitosis itself is divided into five phases: prophase, prometaphase, metaphase, anaphase, and telophase. By the end of telophase, another process called cytokinesis occurs. In Prophase, the chromatin in the nucleus of the cell begins to condense into chromosomes. In prometaphase, the nuclear membrane degrades and microtubules enter the nuclear space. In metaphase, the chromosomes are moved to and aligned in the center of the nucleus by microtubules. In anaphase, the chromosomes are separated by the shortening of microtubules, and in telophase the two nuclei become completely separate and the genetic material begins to de-condense. At this point, cytokinesis has already begun to separate the rest of the cell material by folding in to form a cleavage furrow. When cytokinesis is complete, two cells with identical DNA (daughter cells) are formed from the original cell (DeSouza & Osmani, 2007).

Cell Death
There are two major forms of cell death, apoptosis and necrosis. Apoptosis is programmed cell death (or "cell suicide), while necrosis is cell death triggered by some traumatic damage to the cell. Apoptosis is a multistep cell

signaling process that can occur due to either extracellular or intracellular conditions. Extracellular conditions include the presence of toxins, hormones, or growth factors that trigger the process of apoptosis. Intracellular apoptotic signals include evidence of nutrient deficiency, heat, or radiation. In apoptosis, the cell shrinks and the nucleus condenses as the genetic material fragments. The membrane then breaks the cell into several small vesicles containing organelles, DNA, etc., a process called "blebbing." These "blebs" are then engulfed by phagocytes (cell eating).

Necrosis is premature death of a cell caused by traumatic injury. In contrast to apoptosis, this type of cell death is unnatural, caused by extracellular factors such as toxins, infection, etc. In necrosis, the dead cells are not engulfed by other cells, and therefore must be removed, often by surgical processes, as they may form masses of dead tissue (Karp, 1999).

Cancer

Cancer is the second leading cause of death in Americans. An approximated 1.5 million people were diagnosed with cancer just this year, and about 1,500 die from cancer each day. The most deadly form of cancer in both males and females is lung and bronchus cancer, followed by prostate cancer in males and breast cancer in females (American Cancer Society, 2010).

There are many ways to diagnose and stage cancer. One of the more widely used is the TNM staging system. It looks first at the extent, or size, of the primary tumor (T), then at any lymph node involvement (N), and finally at the absence or presence of distant metastases (M). When T, N, and M are determined, a stage from I to IV is assigned based on level of growth with IV being the most developed (American Cancer Society, 2010).

Cancers are classified by the type of tissue in which they originate and by the primary site of the cancer. There are hundreds of types of cancers, all grouped into six major categories: carcinomas are malignant neoplasms, or tumors, that originate in epithelial tissue and account for 80-90% of diagnoses; sarcomas are cancers that originate in connective and supportive tissues such as bone and tendons; myelomas originate in the plasma cells of the bone marrow; leukemias, or cancers of the bone marrow, are "liquid cancers" and are often associated with the overproduction of white blood cells; lymphomas originate in the lymph nodes; mixed types are any combination of the previous five categories (National Cancer Institute, 2010).

Tumors

Malfunctions in the cell cycle may cause the formation of extra cells, which in turn may develop into a mass of tissue called a growth or tumor. Tumors may be malignant or benign. Benign tumors are not cancer. According to the National Cancer Institute, they are easily removed and rarely life-threatening. Benign tumors are simply growths of cells that will not invade surrounding tissues. Malignant tumors, however, are cancer. They may be life threatening and can grow back after attempted removal. Malignant tumors invade and damage surrounding tissues and may spread to other parts of the body in a process called metastasis. When tumors metastasize, the cells break away from the original (primary) tumor, spread through the bloodstream of lymphatic system, and potentially form new tumors in other organs. At this point the tumors are potentially lethal, because they can no longer be effectively surgically removed (National Cancer Institute, 2006).

Cancer is often divided into two types, solid and liquid cancers. A solid cancer is also called a solid tumor, and is defined by the National Cancer Institute as "an abnormal mass of tissue that usually does not contain cysts or liquid areas." Solid tumors are those that affect organs and tissues, and may be malignant or benign. They are named for the type of cells that form them. Examples of solid tumors are lung and cervical cancer (US National Institutes of Health). Liquid cancers, or tumors, are cancers of the blood or lymphatic system. They include all leukemias and lymphomas (Saporito, 2008).

Causes
Cancer may be caused by both environmental and genetic factors. Environmental causes of cancer may include, but are not limited to, smoking, lack of proper diet and exercise,

alcohol, UV radiation, some viruses or infectious agents, ionizing agents, pesticides, and certain medicines. While this is a long list of fairly common elements of American life, it is very important to know that not everything causes cancer (National Cancer Institute, 2003).

Most cancers are caused by changes in genes. While cancer itself is not passed down between generations of families, it is possible that a certain gene change that makes a person more susceptible to cancer may be passed down from parent to child. For example, breast and colon cancers sometimes run in families. However, the majority of cancers arise from DNA changes in somatic cells over the course of one's life rather than DNA changes in a parent's cells that are passed down (National Cancer Institute, 2006).

Mechanisms

Most cancers result from multiple mutations in genes responsible for growth, growth factor receptors, and intracellular molecules for signaling. Proto-Oncogenes are normal genes that code for proteins and stimulate cell growth. Oncogenes are mutated proto-oncogenes that can result in cancer. Often targeted are tumor-suppressor genes, which inhibit abnormal cell division. 75-98% of all cancers are caused by a mutated p53 gene, which codes for a tumor-suppressor protein. Many others are caused by mutations in the *Ras* protein, which is responsible for the G-protein which relays signals from growth factors to protein kinases during cell signaling.

More than one somatic mutation is necessary before a cell becomes cancerous; usually at least one active oncogene as well as the mutation or loss of multiple tumor-suppressor genes is involved. This causes the loss of cell cycle control in cancer cells: they will no longer respond to density-dependent signals telling them when to stop dividing. The resulting uncontrolled cell growth forms potentially cancerous tumors. The loss of regulators also eliminates anchorage dependence; cells will no longer need to be anchored to a somatic surface to divide, and will therefore be able to metastasize.

The difference between cancer and normal somatic cells is that cancer cells do not have growth inhibition triggers and do not exhibit anchorage dependence or density dependence inhibition, allowing them to divide uncontrollably.

Symptoms

Because of the effects cancer has on cells, its effect on the body is also profound, causing several symptoms that often include thickening or lumps in any part of body, obvious change in a wart or mole, a sore that does not heal, a nagging cough or hoarseness, changes in bowel or bladder habits, indigestion or difficulty swallowing, unexplained changes in weight, and unusual bleeding or discharge (National Cancer Institute, 2006).

Cell lines

The term "cell lines" in this case refers to cancer cells that are able to continue dividing without damage to DNA, and are therefore biologically immortal. The normal methods of cell death and growth inhibition do not function properly, if at all, in these cells.

HeLa

The HeLa cell line is one of the oldest and most commonly used cell lines in science. It is a cervical cancer cell line derived from the cells of Henrietta Lacks, who eventually died from cancer in 1951. The cells are remarkably long-lasting and durable, and were propagated by George Otto Gey at the time of Lacks' death. They were the first human cells to be successfully cultured *in vitro* and have since been used for the testing of the first polio vaccine, cancer research, AIDS research, and more than 60,000 other experiments (Skloot, 2010).

Jurkat T

The Jurkat T cell line was originally established in the 1970's from the blood of a 14-year-old boy with T cell (a type of white blood cell) leukemia. A study took samples from sixty-two leukemia patients. Of those patients, only eight permanently growing cell lines were obtained, and only one of the eight permanently growing lines expressed T cell characteristics and competent receptors. They are sometimes used to study HIV, but the primary purpose of Jurkat T cells is to study the susceptibility of cancers to drugs and radiation (Schneider, Schwenk, & Bomkamm, 1977).

Tree Structure

Plants, specifically trees, are comprised of a variety of organs, tissues, and cells. The three main organs of a plant are the roots, stem, and leaves. The bark is considered to be part of the stem, or trunk, of the tree. Bark makes up the outer covering of woody branches and stems. The bark includes all the plant tissues inside of the cambium, the layer of cells responsible for increased girth in trees. Bark formation starts by cell division in the cambium, which produces xylem (water-carrying tissue) on the inside and phloem (nutrient-carrying tissue) on the exterior bark side. The inner bark of a tree is usually thin and light-colored. When the outer bark becomes warn down through weather and age, the inner layer of bark gradually becomes the outer layer, and subsequently turns to a darker color. The scaly, outermost part of bark on mature trees is known as the rhytidome. Bark is a complex, heterogeneous material that is comprised of a thin, physiologically inactive layer and a complex, inert outer layer, whose main function is to prevent water loss and protect the cambium (Harkin, & Rowe, 1990).

Extracts

The bark extracts that will be tested in this experiment are graviola, cinnamon, black cherry, lapacho, and paper birch. All extracts can be found in their cells' vacuoles or cell walls.

All are said to possess compounds that inhibit the growth and or development of cancer cells.

Graviola

The graviola tree, scientifically known as *Annona muricata*, is a small, upright evergreen tree that grows 5 to 6 meters in height with large, dark green leaves. Graviola trees are indigenous to warm, tropical areas in North and South America, including the Amazon Rainforest. Graviola extract contains annonaceous acetogenins, or phytochemicals (Smith, 05).

These phytochemicals are potent antitumor agents. Regarding their chemical structure, they are long-chain fatty acids. According to a lab study, annonaceous acetogenins reduce levels of ATP by inhibiting complex I of mitochondria as well as NADH oxidase in the plasma membranes of tumor cells. In turn, this inhibition stops ATP-driven processes from occurring in cancer cells (Alali, Liu, & McLaughlin, 1999).

Cinnamon

Cinnamon trees, or *Cinnamomum verum*, are naturally found in tropical rainforests in Sri Lank and southern India. They have also been introduced to Fiji, America Samoa, Cook Islands, Hawaii, Seychelles, and Samoa. This tree is widely cultivated for its bark, which produces cinnamon. This cinnamon tree is commonly known as Ceylon cinnamon, and its bark contains the essential oils cinnamic aldehyde and cinnamyl aldehyde. In a recent laboratory study, it was reported that cinnamic aldehyde and cinnamyl aldehyde have anti-cancer effects associated with modulation of angiogenesis and effector function of CD8+ T-cells. In the study, cinnamic aldehyde and cinnamyl aldehyde induced cellular apoptosis by inhibiting the activities of NFkappaB and AP1 in mouse melanoma cells (Kwon, Hwang, So, Lee, & Sahoo, 2010).

Black (Wild) Cherry

The black cherry tree, or *Prunus serotina*, is native from Canada to the Eastern United States. It is a deciduous tree with white, fragrant flowers. Black cherry bark has a gray-brown color, and is scaly (Pruero). The bark contains anthocyanins and astringents, which are both said to have anti-inflammatory and anti-cancer effects (Prior, & Wu, 2006). Wild cherry extract is said to decrease cyclin D1 expression and increase NAG-1 expression in SW480 and HCT-119 cell lines. Wild cherry extract suppressed beta-catenin/T-cell factor transcription, by recuding cyclin D1 expression. According to data collected in a lab study, this extract induced apoptosis by enhancing NAG-1 expression and decreasing regulation of beta-catenin signaling, which in turn reduced cylin D1 expression in human colorectal cancer cells (Yamaguchi, Liggett, & Baek, 2006)

Lapacho

The Lapacho tree, or *Tabebuia avellanedae,* is native to South America. It is an evergreen tree that grows in warm climates. When the Lapacho tree is mature, it can be as tall as 38.1 meters. This tree also grows characteristic violet-colored flowers. Due to increasing demand for this tree, it is now considered endangered. The lapacho tree contains chemical compounds called naphthoquinones, specifically beta-lapachone and lapachol. Both compounds are said to have antifungal, antiviral, and antibacterial properties. In a laboratory study, beta-lapachone was used on endothelial cells in vitro. After treating endothelial cells with beta-lapachone and lapachol, mitochondrial membrane potential decreased and cGMP levels decreased. During beta-lapachone-induced cell death, caspases and caplain were activated. The data collected during this laboratory study suggested that beta-lapachone has a potential of being an anti-angiongenic drug (Kung, Chien, Chau, Don, & Lu, 2007).

In another study, it was concluded that beta-lapachone induces apoptosis in HepG2 hepatoma cell line through induction of Bax and activation of caspase. Beta-lapachone was shown to inhibit the viability of HepG2 by inducing apoptosis, because DNA fragmentation took place and apoptotic bodies were present. Beta-lapachone-induced apoptosis was associated with proteolytic activation of caspase-3 and-9 and degradation of poly(ADP-ribose) polymerase protein. This study indicated that beta-lapachone has the potential to be a chemopreventive agent for liver cancer (Woo, Park, Rhu, Lee, & Choi, 2006).

Paper Birch

The final bark extract that will be tested is Paper birch, or *Batula papyrifera.* Paper birch can be found across North America, from Alaska to Newfoundland. It is considered to be a medium-sized tree, with a slender trunk. Its bark is thin, white, and smooth on mature trees. The bark is easily peeled off in sheet ("Paper birch, betula," 2006). The bark of paper birch trees contain anticancer diarylheptanoid glycosides in their inner bark (Mshvildadze, Legault, Lavoie, & Gauthier , 2007).

The Pacific Yew Tree and Taxol

The Pacific Yew Tree, scientifically known as *Taxus brevifolia*, is a conifer native to the American Pacific Northwest. It is most commonly found from south to southern California to southern Alaska. The Pacific Yew Tree grows to 10 to 15 meters at maturity, and grows very slowly. Its bark is scaly, thin, and brown, with its outermost scales being purplish, and its inner scales a reddish color. The wood form this tree is very strong. Bark from the Pacific Yew contains taxol, a compound used in the drug, Taxol, which treats various forms of cancer (Taylor, & Taylor). Taxol's mechanism of antitumor activity is unique because it promotes microtubule assembly and stabilizes the microtubules, which prevents mitosis (Huizing et al., 1995). Taxol does this by binding to the B subunit of tubulin, forming microtubule polymers leading to growth arrest in the G2/M phase of the cell cycle (Gotaskie and Andreassi, 1994). Due to taxol's antimitotic activity, it is a useful cytotoxic drug in treating obstinate tumors including head and neck cancer, small cell and non-small cell lung cancer and most notably breast and ovarian carcinomas. It may also slow the course of melanoma. Response rates to taxol treatment varies among cancers. In certain cases, Taxol has also been shown to produce complete tumor remission (Guchelaar et al., 1994).

Extracts vs. Whole Bark

There are many reasons why whole bark is an ineffective and unreliable method of medical treatment. The first reason is that some drugs must be taken intravenously due to the enzymes in the digestive system. These enzymes would inactivate the active chemicals in these drugs, rendering them useless if ingested (Kroll, 2008).

Another reason against the use of bark as treatment is the interference of the composition of the bark. Other compounds found in the bark could influence dissolving and the absorption of the drug, possibly causing inhibiting reactions or enhancing metabolism (Kroll, 2008).

The third reason why using whole bark is not feasible is the low concentration of the desired molecules in the bark itself. In order to take the recommended dosage, a very large amount of bark would need to be ingested (Kroll, 2008).

The fourth major reason is that the amount of the desired molecule varies greatly between bark samples. Therefore, the dosage prescribed when prescribing bark would be neither consistent nor reliable (Kroll, 2008).

A fifth reason is that there may be undesired substances in the bark that were not anticipated to be present. These substances, such as heavy metals and pesticides can lead to harsh side effects (Kroll, 2008).

For these reasons, a purified extract is a more reliable and constant method for treatment than the use of whole bark.

Aqueous vs. Organic Solvents

In an aqueous solution, water is the solvent (Campbell, Reece, & Urry, 2005). If the substance desired is soluble to water, then an aqueous solvent can be used in the extraction process. If water is added to a soluble substance, the resulting mixture can be filtered through a water-permeable medium. The water can then be removed from the remaining mixture, leaving the desired substance (LeMAB, 2010).

However, not all substances are soluble to water. In these cases, another type of solvent such as an organic solvent would be needed for the extraction process. An organic solvent is a carbon-containing solvent, usually liquid, that is used to dissolve substances After dissolving the substance, the new mixture can be filtered and evaporated, leaving the desired substance (DG Health and Consumers, 2009).

Aqueous buffers can also be used in extraction processes. Using an aqueous buffer instead of an organic solvent has benefits including faster extraction and production of extracts that are free of pigment and alcohol-soluble materials. Aqueous buffers differ from aqueous solvents in that they affect the pH of a substance, attempting to keep it at a relatively constant pH. (Jones, 1968)

Centrifugation

Centrifugation is a mechanical process used to separate specific materials from a sample. This process utilizes differences in mass to separate these materials. Using a spinning mechanism, a strong centripetal force is created, resulting in larger, more dense particles to form a pellet at the bottom of the centrifuge tube. This creation of a centripetal force is required because the force of gravity alone is not strong enough to separate the desired materials and another mass/density dependent force was needed. To illustrate the intensity of the force created by a centrifuge: two 30mL tubes of water spinning at 50,000 rpms create a force equal to the force of a 13,000lb truck on Earth due to gravity. (Woodbury, 2000).

Flow Cytometry

Flow cytometry is the process used to characterize cells at very fast rates using a laser beam. This method can be used to determine both physical and chemical properties in the cell. Information gathered by flow cytometry includes whether or not the cell is undergoing apoptosis (programmed cell death), cell type, DNA content, enzyme activity, proteins within the cell, antigens on the surface of the cell, cytoplasm granularity, integrity of the cell membrane, signal transduction (presence of Ca++), DNA synthesis, cell surface receptors, cytokines, oxidative metabolism, pH, RNA, and cell size. In this process, cells are focused into a stream of single cells. When the light source (the laser beam) hits the cells, it scatters and excites fluorescent markers in the cells. Differences in cell cycle stage, activation, etc. result in morphological changes in the cell; these morphological changes cause a different refraction of light in both the forward light scatter and the side light scatter. This difference can be detected by sensors and analyzed by a computer to provide a great amount of information about the cells in a very short period of time. This process can be used to determine the characteristics of cells at a rate over 20,000 times faster than by hand (University of Texas, 2006).

Cisplatin

Cisplatin (also referred to as DDP, cisplatinum, and cis-diamminedichloridoplatinum(II) or CDDP) is a platinum-based chemotherapy drug. Cisplatin functions by the platinum complexes interfering with the cells' DNA, resulting in severe DNA damage. More precisely, the molecule covalently bonds with ligands including water. The molecule then loses it's chlorine atom and receives a hydroxyl groups, and eventually a part of the DNA. This occurs through alkylating. Through this process, alkylating agents form electrophiles which seek out nucleophilic atoms and functional groups. Guanine, a nucleotide in DNA, is affected by this process. The cisplatin binds on one side to Guanine and another to a protein, preventing it from being removed by repair methods of the cell. The cell then attempts to repair this damage, but when it realizes that the DNA is unable to be repaired, it undergoes apoptosis, or programmed cell death. This drug has been used to decades alone or in combination with other cancer treatment drugs such as vinorelbine, ifosfamide, and gemcitabine to treat cancers including non-small-cell lung cancer, melanoma, liver cancer and cervical cancer (Jones, 2006-2010).

Due to the fact that cisplatin interferes with DNA in order to cause the cell to undergo apoptosis, this drug can also cause secondary cancers. If the cell's DNA is damaged by cisplatin without undergoing cell death, mutations occur resulting in malignant cells. In most cases, the benefits of using this drug outweigh the risks that accompany it (Jones, 2006-2010).

Unlike cisplatin, transplatin (an isomer of cisplatin) does not function as a cancer treatment drug. This is due to the different location of chlorine atoms, which do not cause cross-linking (the method by which cisplatin works) to occur (Jones, 2006-2010).

Cisplatin is administered intravenously in a dissolved saline solution at a rate of 1mg/minute infusion. The dosage and schedule of infusion is dependent on the type and severity of the cancer (Jones, 2006-2010).

Chemotherapy (general)

Chemotherapy is the use of drugs to treat cancer by killing cancer cells. This treatment, administered either intravenously or orally, damages both non-cancerous and cancerous cells. This results in overall weakness, digestive issues, alopecia (loss of hair), and decreased fertility. Damage of many vital cells in the body such as those involved in the immune system can also result in susceptibility to infection. Symptoms are mostly temporary and can be treated with the help of a doctor; therefore chemotherapy is seen as an effective and worthwhile cancer treatment, regardless of the side effects (National Cancer Institute, 2006).

Cell Culturing

Cell culturing is the process of maintaining a line of cells by providing those cells with the proper nutrients and conditions. These cells are grown *in vitro*, or in a glass (usually a flask or Petri dish). Cells cultured can either be a genetically identical population of clones or contain genetic variation. These cells can be cultured in order to study biology on a cellular level, to test the effects of drugs on human cells without the need of human subjects, and to attempt to make other scientific advances (Chaundry, 2004).

Conclusion

The goal of this experiment is to test the effects of tree extracts to determine whether or not they will inhibit growth or cause death of cancer cells when in conjunction with and in place of the chemotherapy drug cisplatin. If conclusive results are found, a method of chemotherapy could possibly be developed using these plant extract.

Hypothesis

Due to the research conducted, it is hypothesized that bark extracts will be found to have biologically active compounds that inhibit growth of or kill cancer cells, and that the graviola extract will be the most effective in killing or inhibiting growth of cancer cells..

It is also hypothesized that the extracts will not be more effective than Cisplatin, but will act as an effective complement, causing higher percentages of cell death when the chemotherapy drug is combined with extracts and applied to cells.

Procedure
Preparing the Bark

1. Acquired 1 lb of *Cinnamomum verum* (cinnamon)

bark, 3 small logs of *Betula papyrifera* (paper birch), 1 lb of *Tabebuia Avellaneda* (lapacho) bark, 1 lb of *Prunus serotina* (black cherry) bark, and 2 fl. oz of *Annona muricata* (graviola) extract.
2. Used knife to remove inner and outer bark from the paper birch logs. Kept inner bark.
3. Dried all barks for 1 week.
4. Ground barks in food processor/coffee grinder until fine powder.
5. Stored in sealable Tupperware containers.

Making PBS and Methanol Extracts (the following do not include graviola extract)

1. Labeled 3 tubes for each bark, 1 for each solvent used.
2. Measured out 1g of each of the bark powders using weigh boats scoopula and electronic balance.
3. Pipetted 49.5 mL of Methanol in to 50mL test tube using pipet-aid.
4. Added .5 mL of acetic acid to the test tube.
5. Covered and inverted to mix. (1% acetic acid, 99% methanol) Vortexed.
6. Added 1 mL of PBS to each bark powder labeled PBS. This wasn't enough, so 4 mL more of PBS were added (5mL total). Vortexed.
7. Put in water at 90° C to help dissolve. Shook every few minutes.

8. Added 5 mL of acidified methanol solution to each powder labeled methanol.
9. Sonicated solutions at 50% for 10 seconds each, making sure the sonicator tip didn't touch the side of the tube.
    a. Sonicated methanol and PBS once, excepting cherry, which was sonicated 3 times waiting at least 1 minute between each time.
10. Centifuged solutions at 1600 rpm for 5 minutes.
11. Labeled 3 new tubes for each extract with bark type, solvent, and "supernatant"
12. Removed 2 mL of supernatant from the centrifuged acidified methanol tubes using 1 mL pipette, expelling into the corresponding supernatant tubes.
13. Removed as much supernatant as possible from the centrifuged PBS tubes using 1 mL pipette, expelling into the corresponding supernatant tubes.
14. Added hexane to remaining PBS extract (not supernatant). Two phases formed, and the tube was thus discarded.

Sterilizing Extracts

1. Sterile Technique
    a) Hands must be kept in the hood
    b) Hands may not pass over anything
    c) Hands may not touch any non-sterile objects
    d) Rinse surface and hands with 70% Ethanol
2. Using Sterile Technique:

a) Labeled sterile tubes with extract, solvent and "sterile."
   b) Removed cap from supernatant liquid.
   c) Held sterile tube at an angle, remove cap and place on table (on side). Returned tube to rack.
   d) Opened filter package without passing thumb passing over. Disposed packaging in trash.
   e) Opened syringe package. Removed plunger. Placed on table. Removed syringe cap. Disposed packaging in trash.
   f) Screwed syringe onto filter.
   g) Capped a 1 mL pipette tightly. Drew 1 mL liquid, expelled into syringe. Disposed of pipette tip.
   h) Held syringe over sterile tube labeled for filtered extract. Expelled liquid into tube by placing plunger into syringe.
   i) Replaced cap at an angle.
3. Repeated for all PBS and methanol extracts.

Making DMSO Extracts

1. Labeled 1 tube for each bark type plus DMSO.
2. Measured out 1g of bark powder into corresponding tubes.
3. Added 3mL of DMSO to each extract using 1 mL pipette. Vortexed.

4. Changed pipette tips between extracts
5. Placed in centrifuge, set to 1600rpm. Let run for 5 minutes.
6. Supernatant did not form so 2 mL more of DMSO were added to each tube.
7. Extracts were put on the rocker for 10 minutes.
8. Centrifuged DMSO extracts for 5 min at 1600 rpm.
9. Used sterile technique and 1 mL pipette to expel 1 mL of supernatant through nylon filter into 1.5 mL microfuge tubes.

Plating Cells: Jurkat T and HeLa viability

1. Two Jurkat plates were preset with 1 mL of Jurkat cells and one mL of media in each well. HeLa cell plates were preset with 500 µL of HeLa cells and 500 µL of cell culture media in each well.
2. Used sterile technique to add 10 µL of extract or solvent to the corresponding wells and plates.
    a. Changed pipette tips between wells.
    b. Swirled plate after addition of extract/solvent to each well.

*The Cisplatin was dissolved in 9% NaCl solution at concentration s of 2 µM and 20 µM, and added by Tom.

*Due to time constraints, 1 Jurkat plate and the HeLa cell plates were prepared by Fay, a graduate student in the lab.

3. Incubated for 16 hours.

Preparing Staining Solution

1. Used Pipet-aid to draw up 14 mL of 1-X PBS, expel into test tube.
2. Used 20 µL micropipette to draw up 14 µL of PI (propidium iodide), expelled into test tube.

Preparing Jurkat cells for Viability Test

1. Observed Jurkat cells in microscope.
2. Labeled test tubes, one for each well of one Jurkat plate, and one unstained.
3. Used a 1 mL pipette to draw up and expel 100 µL of cells twice in well to mix, then drew up 100 µL of cells. Expelled into labeled test tubes.
4. Added 400 µL PI solution to each test tube, except the unstained trial.
5. Add tubes to ice bucket.
6. Ran through flow cytometer.

Using Flow Cytometer

1. Set to record 10,000 events.
2. Tapped cells to mix.
3. Selected corresponding trial on the computer.

4. Brought the base all the way to the left.
5. Put tubes on the needle.
6. Returned base to the right.
7. Clicked "acquire data."
8. After 10 seconds, clicked "record data."
9. When data collection was complete, clicked "remove tube."
10. Brought the base all the way to the left and removed the tube.
11. Waited for flush.
12. Repeated steps 2-10.

Preparing HeLa Cells for Viability test

1. Observed HeLa cells in microscope.
2. Trimmed "rubber policeman" scrapers to the size of the well plates.
3. Used scrapers to remove HeLa cells from the bottom of the plates.
4. Pipetted cells into centrifuge tubes using 3-6-9-12 method.
5. Centrifuged tubes for 5 minutes at 1600 rpm.
6. Aspirated tubes to remove media.
7. Added 100 µL buffer solution of trypsin buffer solution.
8. After 3 minutes, added 100 µL of tissue culture media with fetal calf serum and EDTA+ cations.
9. Added 400 µL PBS and PI solution.

10. Used 1 mL pipette to transfer cells from test tubes to microfuge tubes.
11. Microfuged tubes on 3 for 3 minutes.

Running HeLa Cells in the Flow Cytometer

1. Labeled new test tubes for each well.
2. Used a 1 mL pipet to draw up 300 μL of cells from the microfuge tubes. Expelled into appropriate test tube.
3. Ran in Flow Cytometer using outlined method.

Cell Cycle Analysis

1. Labeled microfuge tubes for eight trials:
   a. PBS control, graviola extract, cisplatin (20 μM), cells alone, Wild cherry- PBS, Lapacho- PBS, Birch- PBS, and cinnamon- PBS.
2. Prepared staining solution for cell cycle analysis.
   - 500 μL 2x-PBS + .2% Triton x-100
   - 400 μL distilled water
   - 50 μL Rnase 3 A stock
   - 50 μL PI stock
3. Added 1 mL Jurkat cells and 500 μL PBS were added to microfuge tubes.
4. Spun in microfuge on 3 for 3 minutes.
5. Aspirated microfuge tubes.
6. Added 50 μL staining solution to Jurkat pellet.

7. Incubated in 37.2°C water bath for 30 minutes.
8. Ran through flow cytometer using previously outlined method.

   *Cells for cell cycle anaylsis taken from from Fay's Jurkat plate.

Annexin-V

1. Labeled microfuge tubes for eight trials:
   a. PBS control, graviola extract, cells-unstained, cells alone, Wild cherry- PBS, Lapacho- PBS, Birch- PBS, and cinnamon- PBS.
2. Diluted Annexin-V buffer from 10x to 1x by adding 5 mL of buffer to 45 mL of distilled water.
3. Added 1mL of Jurkat cells to microfuge tubes.
4. Cells washed with 400 µL PBS.
5. Microfuged cells on "3" for 3 minutes.
6. Aspirated liquid and scraped tube against a test tube rack to dislodge pellet.
7. Resuspended in 500 µL of 1x Annexin buffer.
8. Put on ice.
9. Added 5 µL Annexin-V in dark.
10. Wrapped test tube rack in foil. Let sit for 15 minutes.
11. Ran Annexin-V trials through flow cytometer with previously outlined method.

    * Cells for Annexin-V taken from Fay's Jurkat plate.

Caspase-3

1. Labeled microfuge tubes for eight trials:
    a. PBS control, graviola extract, cells-unstained, UV light, Wild cherry- PBS, Lapacho- PBS, Birch- PBS, and cinnamon- PBS.
2. Diluted Caspase-3 buffer from 10x to 1x by adding 5 mL of buffer to 45 mL of **distilled water**.
3. Added 1mL of Jurkat cells to microfuge tubes.
4. Cells washed with 400 µL PBS.
5. Microfuged cells on "3" for 3 minutes.
6. Aspirated liquid and scraped tube against a test tube rack to dislodge pellet.
7. Resuspended in 500 µL Caspase buffer.
8. Put on ice and stored in freezer for 16 hours.
9. Exposed 3 mL of Jurkat cells and media in small petri dish to UV light for 10 minutes.
10. Added 3 mL to another petri dish. Let both sit for 1 hour.
11. Used a micropipette to mix cells in well plate using "12, 3, 6, 9" method.
12. Prepared and labeled two microfuge tubes: UV light and just cells
13. Added 1 mL of normal cells to respective microfuge tube using 1 mL pipette.
14. Added 500 µL 1x-PBS.
15. Spun all tubes in microfuge on "3" for 3 minutes.
16. Aspirated microfuge tubes.
17. Flicked to dislodge pellet.

18. Added 1 mL of Tritan x-100 buffer to each.
19. Let sit for 20 minutes.
20. Spun down tubes on "3" for 3 minutes.
21. Resuspended in 1 mL of a fixative buffer containing formaldehyde at 1%.
22. Spun down on "3" for 3 minutes and aspirated tubes.
23. Added 1 mL PBS.
24. Spun down on "3" for 3 minutes.
25. Pipetted of 1 mL and added 200 µL PBS.
26. Ran through flow cytometer using method previously outlined method.

Jurkat- Cisplatin with Extracts

1. Observed cells under microscope.
2. Used a pipet-aid and sterile technique to add 1 mL of Jurkat cells in media to 20 wells in a 24 well plate.
3. Used a 20 µL micropipette to add 1 µL of each extract to its corresponding well.
4. 100 µM Cisplatin was added by Tom.
5. Incubated for 16 hours.
6. Made the PI stain in dark. Put on ice.
   - 10 mL PBS
   - 10 µL PI
7. Labeled 17 tubes.
8. Drew up 100 µL of cells then expelled twice to mix.
9. Drew up 100 µL of cells and expelled into

corresponding test tube.
10. Repeated steps 7 and 8 for each well, changing pipet tips in between.
11. Used 1 mL pipette, add 400 µL of PI staining solution to each tube. Stored on ice.
12. Run through flow cytometer using previously outlined method.

Data

HeLa/Extract/Solvent

Variable	Alive	Dead
Untreated	76	24
PBS Control	16.4	83.6
Acetic Methanol	22.5	77.5
DMSO	49.6	50.4
Cisplatin 2	2.3	97.7
Cisplatin 20	5.4	94.6
Cherry PBS	14.9	85.1
Birch PBS	12.8	87.2
Lapacho PBS	2.4	98.4
Cinnamon PBS	43.6	56.3
Cherry Methanol	38.4	61.6
Birch Methanol	14.7	85.3
Lapacho Methanol	48.9	51.1
Cinnamon Methanol	27.8	72.2
Cherry DMSO	59.4	40.6
Birch DMSO	34.5	65.5
Lapacho DMSO	9.9	89.9
Cinnamon DMSO	23.4	76.2
Graviola	0.2	99.8
Ethanol	43.1	56.8

Jurkat/Extract/Solvent

Variable	Cells	Alive	Dead
Unstained	98.3	97.2	2.8
Untreated	97.5	83.6	16.4
PBS Control	98.3	89.9	10.1
Acetic Methanol Control	98.2	87.2	12.8
DMSO Control	98	80	19.9
Cisplatin 2	97.6	82.2	17.8
Cisplatin 20	98.1	79.3	20.7
Cherry PBS	67.6	27.3	72.7
Birch PBS	97.4	73.8	26.2
Lapacho PBS	97.2	67.5	32.5
Cinnamon PBS	94	30.4	69.6
Cherry Methanol	71.4	28.1	71.9
Birch Methanol	97.6	69.4	30.6
Lapacho Methanol	95.7	19.9	80.1
Cinnamon Methanol	90.9	15.7	84.3
Cherry DMSO	89.5	29.8	70.1
Birch DMSO	65.3	52.9	47.1
Lapacho DMSO	50.4	6.6	93.4
Cinnamon DMSO	71.1	28.6	71.4
Graviola	98.4	0.1	99.9
Ethanol	98.2	79.4	20.5

Cell Cycle Analysis

Extract	<G0	G0/G1	S	G2M

Cells	1.3	58.5	28.1	12
PBS	0.7	60.3	27.1	12.4
Cherry	12.2	48.5	23.9	15.6
Birch	0.8	58.1	29.1	12.3
Lapacho	1.6	65.3	21.5	11.5
Cinnamon	14	43.8	27.6	14.6
Graviola	4.9	52.7	27.9	14.4
Cis 20	1.4	57.5	27.8	13.3

Annexin V

Extract	Positive	Negative
Cells	12	88
PBS	14.1	85.9
Cherry	35.6	64.4
Birch	15.8	84.2
Lapacho	19	81
Cinnamon	92.5	17.5
Graviola	93.3	16.7
Unstained	0.1	99.9

Caspase-3 Jurkat

Variable	Positive	Negative
Cells	0.7	99.3
PBS	9.3	90.7

Cherry		19.2	80.8
Birch		15.1	84.9
Lapacho		7.8	92.2
Cinnamon		47.7	52.3
UV (negative control)		28.5	71.5
Graviola		18.1	81.9

Jurkat Extract + Cisplatin Viability

Variable	Cells	Alive	Dead
Unstained	98	100	0
Stained	98.2	93.9	6.1
Cisplatin 20	95.7	17.7	82.3
Birch Only	97.5	91.1	8.9
Birch/Cis	96.9	5	95
Graviola	98.5	93.5	6.5
Graviola/Cis	96.5	7.1	92.9
Lapacho	98.9	94	6
Lapacho/Cis	97.2	6.1	93.9
Cinnamon	98.4	91.6	8.4
Cinnamon/Cis	96.8	5.5	94.5
Cherry	98.8	94.5	5.4
Cherry/Cis	96.7	5.2	94.8

Conclusion

The purpose of this experiment was to determine the effectiveness of tree bark used in conjunction with and in place of the chemotherapy drug Cisplatin (*cis*-diamminedichloroplatinum (II)) to inhibit growth or cause death of cancer cells. This was done through the use of extracts from *Annona muricata* (graviola), *Cinnamomum verum* (cinnamon), *Betula papyrifera* (paper birch), *Prunus serotina* (black cherry), and *Tabebuia avellanedae* (lapacho). The extracts were used in conjunction with cisplatin to treat two types of human tumor cells: HeLa cells and Jurkat T cells.

Cancer is a disease that 1.5 million people are diagnosed with worldwide every year. This disease is a result of mutations in the DNA of cells, causing uncontrolled cell division. This is a result of the cells defying the checkpoints in the cell cycle. If there is a problem in the DNA of a cell, the cell initiates a pathway of chemical reactions that result in the cell's programmed death, also known as apoptosis. This process is beneficial to the organism. Necrosis, the other possible way of cell death, is the result of a different pathway and it is almost always detrimental to the organism. With this process, surgical removal of dead cells is required. Tumors are masses of cancer cells that are dividing uncontrollably. Two commonly used cell lines, which were also used in this experiment, are Jurkat cells (liquid leukemia cells) and HeLa cells (solid cervical cancer cells). Extracts used, as stated above, were chosen due to some, but not a significant amount of research having been conducted concerning their ability to treat cancer cells. These extracts can be compared to the commonly used chemotherapy drug, cisplatin. This drug utilizes its ability to interfere with nucleotides in the cell to induce apoptosis. To measure the effects of these extracts alone and in conjunction with cisplatin, a flow cytometer can be used. This device aspirates cells one-by-one using a fluidic system, allowing each cell to be analyzed individually. A laser beam is projected through the cells, and depending on the light scatter and fluorescence, various characteristics of the cell can be analyzed.

The first step of the experiment was gathering and extracting the barks. Logs were gathered from various places and debarked. These barks were dried and then extracted using phosphate buffer solution (PBS), 1% acetic methanol, and dimethyl sulfoxide (DMSO). Viability tests were conducted on both the Jurkat and HeLa cells to see whether or not the extracts alone would result in cell death. After examining those results, PBS extracts on Jurkat cells were chosen to be used in the rest of the testing due to both the fact that PBS is less harsh on cells and the fact that the data showed high cell death in these trials. Cell cycle analysis was then performed on these cells. Utilizing knowledge of the amount of DNA in different stages of the cell cycle, we can accurately predict where a cell is in the cycle. The cells were then stained with Annexin V to detect the presence of phosphatidylserine, a molecule in the cell's membrane that relocates to the outer membrane during apoptosis. Caspase-3 testing was also carried out, determining whether or not apoptosis was occurring by detecting the activity of the caspase-3 protein. Finally, the Jurkat cells with PBS extracts were tested in conjunction with cisplatin to determine whether or not they synergized, resulting in increased effectiveness.

Unfortunately, the data collected from the HeLa cells could not be analyzed. Because HeLa cells are adherent, they must be separated before data is collected, or the clumps of cells will clog the flow cytometer. To do this, the cells had to be scraped from the bottom of the cell plates and trypsinized, which caused significant cell death in the controls.

Data from the synergy assays was also inconclusive. This is because, due to time constraints, a dose titration of cisplatin (used to find the concentration that will kill about 50% of the Jurkat T cells) could not be performed. As a result, the concentration of cisplatin used caused too much cell death, and there was not enough variation between results from the extracts alone to draw conclusions.

Though some assays were inclusive, the data showed that the hypothesis was correct with respect to the effectiveness of tree bark extracts in killing cancer cells. It was also correct that the Graviola extract would be the most effective. Graviola was not, however, followed by lapacho in effectiveness, but rather cinnamon. The data collected is outlined in the following paragraphs.

Wild Cherry extracts exhibited high death (PBS: 72.7% dead, methanol: 71.9% dead, DMSO: 70.1% dead) in the viability assay, indicating the presence of bioactive compounds. The cell cycle analysis showed a relatively high percentage,

12.2%, of cells in the phase <G_0, indicating that the cells may be undergoing apoptosis. A large percentage of cells, 35.6% were positive for Annexin V staining, which reaffirmed the high cell death from PI staining. Caspase-3 tests also showed a high percentage of cells, 19.2%, staining positive, indicating active caspase-3 was present in those cells and they were therefore undergoing apoptosis. While the active compound in cherry was not determined through this series of experiments, research shows that it may actually be a group of compounds called anthocyanins.

Paper Birch extracts exhibited low death in the viability assay, excluding the DMSO extract (PBS: 26.2% dead, methanol: 30.6% dead, DMSO: 47.1% dead), indicating the presence of bioactive compounds in the DMSO extract only. The cell cycle analysis showed normal percentages of cells in the phase <G_0, 0.8%, indicating that the cells most likely were not undergoing apoptosis. A relatively low percentage of cells, 15.8%, were positive for Annexin V staining, which reaffirmed the low cell death from PI staining. Caspase-3 tests showed a high percentage of cells staining positive, 15.1%, indicating active caspase-3 was present in those cells and they were therefore undergoing apoptosis. However, cell death was still low. While the active compound in birch was not determined, research shows that it may be a compound called betulin, or a group of compounds named diarylheptanoid glycosides, which are also found in some plants used in traditional Chinese medicine.

Lapacho extracts exhibited low death in the viability assay (PBS: 32.5%, Methanol: 80.1%, DMSO: 93.4%), except the DMSO extract, indicating the presence of bioactive compounds in the DMSO extract only. The cell cycle analysis showed normal percentages of cells in the phase <G_0, 1.6%, indicating that the cells most likely were not undergoing apoptosis. There was, however, a relatively large percentage of cells in the G_0/G_1 phase, indicating an abnormality in the cell cycle. A relatively low percentage of cells, 19%, stained positive for Annexin V, which reaffirmed the low cell death from PI staining. Caspase-3 tests also showed a low percentage of cells staining positive, 7.8%, indicating that active caspase-3 was not present in those cells and therefore they were not undergoing apoptosis. The active compound in lapacho was not determined, but research shows that it may be either of the compounds, napthaquinones or anthraquinones, both of which are soluble in solution of large, slightly polar organic molecules such as DMSO, accounting for the high cell death only in DMSO extracts.

Cinnamon extracts exhibited high death in the viability assay (PBS: 69.6%, Methanol: 84.3%, DMSO: 71.4%), indicating the presence of bioactive compounds. The cell cycle analysis showed a higher than normal percentage of cells in the phase <G_0, 14%, indicating that the cells may be undergoing apoptosis. A large percentage of cells, 92.5%, were positive for Annexin V staining, reaffirming the high cell death from

PI staining. Cinnamon had the highest percentage of cells stain positive for Caspase-3, 47.7%, indicating active caspase-3 was present in those cells and they were therefore undergoing apoptosis. The active compound in cinnamon was not determined through this series of experiments, but research shows that it may be cinnamyl or cinnamic aldehyde.

Graviola extracts exhibited high death in the viability assay (99.9% dead), indicating the presence of bioactive compounds. The cell cycle analysis showed a higher than normal percentage of cells in the phase <G_0, 4.9%, indicating that the cells may be undergoing apoptosis. A large percentage of cells were positive for Annexin V staining, 93.3%, which reaffirmed the high cell death from PI staining. Caspase-3 tests also showed a high percentage of cells staining positive, 18.1%, indicating that active caspase-3 was present in those cells and they were therefore undergoing apoptosis. Research shows that the active compound in graviola may be a group of compounds called annonaceous acetogenins.

The statistical analysis performed was a calculation of CV, or coefficient of variation, which is equal to the standard deviation of a sample divided by the mean. CV is used to compare samples in flow cytometry because of the large amount of variation between samples using different compounds. It supplies a more reliable method of comparison than standard deviation. CV was calculated for each sample in the viability assays, as well as the dead cells and live cells within that sample. In the cell cycle analysis, the CV was calculated for the samples as a whole, and then for each phase of the cell cycle. CV was calculated for each sample as well as the positively stained cells in the Caspase-3 and Annexin V assays.

Coefficient of Variation is used similarly to standard deviation when comparing samples. By calculating the median ± CV for each sample and determining whether or not the intervals overlap, it can be determined whether the cells considered dead are statistically different than those that are considered alive by the flow cytometer. It can also be determined whether two samples are statistically different.

For the Jurkat T cell, HeLa cell, and Synergy viability assays, the median ± CV was used to determine that the alive cells were statistically different from the dead cells; this was the case for each sample, confirming that our data is viable. Median ± CV was also used to determine that the dead cells in each sample with extracts were statistically different than those in the corresponding controls. This was the case for each sample of the three assays. Median ± CV was used in the cell cycle to determine that each phase of the cell cycle was statistically different than the next for each sample, confirming the reliability of the data collected. For Caspase-3 and Annexin V, Median ± CV was used to determine that each sample was statistically different. The statistical analysis supported the conclusion in all cases.

Though statistically our data is viable, there were some errors that occurred during this experiment. Firstly, Birch bark was gathered from a group member's backyard and was not freshly harvested; compounds may have degraded over time. Alternatively, outer bark may have been included rather than inner bark alone. Also, the wild cherry bark was ground less finely than the other barks, possibly decreasing efficiency of extraction and therefore the amount of the biologically active compound(s) as well as the effectiveness of the extract as measured by the flow cytometer. The HeLa cells were attached to the well plates and the process required to remove them (ie- trypsinization) resulted in significant cell death. Lastly, during the Caspase-3 assay a graviola pellet was partially aspirated, and a full 10,000 cell sample could not be analyzed.

There are many possible directions that could be taken with this project in the future. This project provides multiple possible new treatments for leukemia patients, the most promising of which are cinnamon, wild cherry, and graviola bark extracts. However, DMSO extracts of the lapacho and birch barks were not tested further and may also provide a viable option. To simply test the accuracy of the results, the same experiment could be performed with the same species of barks from different sources. This would ensure that the companies providing the bark for this particular experiment do not alter their products in a way that would have amplified our results. Other possible continuations include performing dose titrations of extracts and cisplatin in a synergy assay to optimize effectiveness, using a more carefully defined time course for the Caspase-3 assay to determine the peak of apoptosis, or expanding upon experiments with DMSO or acetic methanol extracts in future experiments rather than focusing on PBS extracts. More long-term continuations include purifying the bioactive compound(s) in extracts and identifying biologically active molecule(s) or compound(s). It would also be possible to determine if bioactive compounds could be used to treat other cancers. The goal is that the synthesis of the compound in the laboratory would be eventually possible so as to increase the sustainability of the drug industry and decrease deforestation. Through this project, it was determined that there are multiple options for new cancer treatments through the extraction of tree barks, specifically those used in this experiment, and there are many possibilities for further experimentation in the future.

References

Alali, F, Liu, X, & McLaughlin, J. (1999). Annonaceous acetogenins: recent progress.

American Cancer Society. (2010). *Cancer Facts & Figures 2010.* Atlanta: American Cancer Society.

Campbell, N. A., Reece, J. B., & Urry, L. (2005). *Biology AP Edition* (7th Edition ed.). San Francisco: Prentice Hall School Group.

Chaundry, A. (2004, August). Cell Culture. *The Science Creative Quarterly* .

DeSouza, C., & Osmani, S. (2007). Mitosis Not Just Open or Closed. *Eukaryotic Cell* , 1521-1527.

DG Health and Consumers. (2009, July 10). *Glossary*. Retrieved November 6, 2010, from Green Facts: http://copublications.greenfacts.org/en/hearing-loss-personal-music-player-mp3/glossary-hearing-loss-personal-music-player-mp3.htm

Gotaskie, G.E. and B.F. Andreassi. 1994. Paclitaxel: a new antimitotic chemotherapeutic agent. Cancer Pract. 2: 27-33.

Guchelaar, H.J., C.H. ten-Napel, E.G. de-Vries, and N.M. Mulder. 1994. Clinical, toxicological and pharmaceutical aspects of the antineoplastic drug taxol: a review. Clinical Oncology -R- Coll Radiology. 6: 40-48

Harkin, J, & Rowe, J. (1990). Bark composition. *Bark and its possible uses* (pp. 2-3). Houston, TX:U.S. Department of Agriculture

Huizing, M.T., V.H. Misser, R.C. Pieters, W.W. ten-Bokkel-Huinink, and C.H. Veenhof. 1995. Taxanes: a new class of antitumor agents. Cancer Invest. 13: 381-404.

Jones, E. G. (2006-2010). *Cisplatin*. Retrieved November 8, 2010, from Cisplatin: http://www.cisplatin.org/

Karp, G. (1999). *Cell and Molecular Biology* (2nd Edition ed.). New York: John Wiley & Sons, Inc.

Kroll, D. J. (2008, November 4). *Why We Don't Prescribe Bark for Cancer*. Retrieved November 8, 2010, from Science-Based Medicine: http://www.sciencebasedmedicine.org/?p=276

Kung, HN, Chien, CL, Chau, GY, Don, MJ, & Lu, KS. (Ed.). (2007). *Involvement of no/cgmp signaling in the apoptotic and anti-angiogenic effects of beta-lapachone on endothelial cells in vitro.*

Kwon, HK, Hwang, JS, So, JS, Lee, CG, & Sahoo, A. (Ed.). (2010). *Cinnamon extract induces tumor cell death through inhibition of nfkappab and ap1.*

LeMAB. (2010). *Aqueous Extraction, Purification and Drying of Extracts*. Retrieved October 28, 2010, from LeMAB: A Solution for Your Extraction Needs: http://www2.ing.puc.cl/lemab/process/aqueous.htMshvildadze, V, Legault, J, Lavoie, S, &

Gauthier, C. (Ed.). (2007). *Anticancer diarylheptanoid glycosides from the inner bark of betula papyrifera.*

National Cancer Institute. (2003, August). Cancer and the Environment. USA: National Institute of Health.

National Cancer Institute. (2006, August). What You Need to Know About Cancer. United States of America: National Institute of Health.

Paper birch, betula papyrifera. (2006, March 04). Retrieved from http://www.rook.org/earl/bwca/nature/trees/betulapap.html.

Prior, RL, & Wu, X. (Ed.). (2006). *Anthocyanins: structural characteristics that result in unique metabolic patterns and biological activities.*

Pruero, P. (n.d.). *Prunus serotina.* Unpublished manuscript, Department of Botany, University of Connecticut, Storrs, Connecticut. Retrieved from **http://www.hort.uconn.edu/plants/p/pruero/pruero1.html**

Sadava, D., Hillis, D. M., Heller, H. C., & Berenmaum, M. R. (2011). *Life: The Science of Biology* (9th edition ed.). Sunderland, MA: Sinauer Associates.

Saporito, B. (2008, September 15). A Foe With Many Faces. *Time Magazine*, pp. 42-43.

Schneider, U., Schwenk, H.-U., & Bomkamm, G. (1977). Characterization of EBV-genome negative "null" and "T" cell lines derived from children with acute lymphoblastic leukemia and leukemic transformed non-Hodgkin lymphoma. *International Journal of Cancer , 19* (5), 621-6.

Skloot, R. (2010). *The Immortal life of Henrietta Lacks.* New York: Crown/Random House.

Smith, J. (05, 17 07). *Graviola extract from the amazon has been found to be stronger than chemotherapy for alternative cancer treatment.* Retrieved from healthyheartht.info/graviola/htm

Smith, J. A., & Martin, L. (1973). Do Cells Cycle? *Proceedings of the National Academy of Science , 70* (4), 1263-1267.

Taylor, R, & Taylor, S. (n.d.). *Taxus brevifolia.* Retrieved from http://www.efloras.org/florataxon.aspx?flora_id=1&taxon_id=233501253

University of Texas. (2006, November 7). *How a Flow Cytometer Operates.* Retrieved November 8, 2010, from Science Park Research Division: http://sciencepark.mdanderson.org/fcores/flow/files/Operation.html

US National Institutes of Health. (n.d.). *Definition of Solid Tumor*. Retrieved November 4, 2010, from National Cancer Institute: http://www.cancer.gov/dictionary/?CdrID=45301

Woo, HJ, Park, KY, Rhu, CH, Lee, WH, & Choi, BT. (Ed.). (2006). *Beta-lapachone, a quinone isolated from tabebuia avellanedae, induces apoptosis in hepg2 hepatoma cell line through induction of bax and activation of caspase..*

Woodbury. (2000, February 15). *Centrifugation*. Retrieved November 1, 2010, from Arizona State University: http://www.public.asu.edu/~laserweb/woodbury/classes/chm467/bioanalytical/centrifugation%20notes.html

Yamaguchi, K, Liggett, JL, & Baek, SJ. (Ed.). (2006). *Antiproliferative effect of horehound leaf and wild cherry bark extracts on human colorectal cancer cells.*

SPF Effectiveness in Preventing UV Damage to DNA Repair-Deficient Yeast

By Caroline Flores and Carrie Goodrich

Hypothesis

If yeast colonies are exposed to artificial UVB light for one minute, then the colonies will have died. Sunscreen use will minimize damage caused by UV light although any SPF beyond 30 will not affect how many colonies are killed any more than SPF 30 would. Thickness of the sunscreen applied will not make a difference in defending the yeast colonies from UV light.

Introduction

Nearly everyone has heard that Ultraviolet, or UV, light is dangerous, but just how dangerous is it and is there a way to prevent the damage it causes? It is complicated and dangerous to answer these questions through experiments involving the human body, however one can still reach conclusions about the damage of exposure to UV light based on its effects on other organisms, such as yeast colonies.

The sun emits energy over a large spectrum of wavelengths. Only about half of the light emitted by the sun is visible to the naked eye. UV radiation is included in the invisible light emitted by the sun. The earth has an ozone layer which blocks most UV radiation, though some of it still reaches the earth. The short wavelengths of UV light can cause damage to the skin, such as sunburn and other health issues, such as skin cancer. This is a large problem connected to UV exposure as it is very common; 1 out of every 5 Americans will develop skin cancer. One's risk of getting skin cancer increases with sun damage; those who experience many blistering sunburns before reaching age 18 face a higher risk of melanoma, the most serious form of skin cancer..

Ultraviolet radiation causes changes in a cell's DNA, thus harming the cell. Most cells are able to repair this damage but there is sometimes a cell that does not. A cell that does not repair this damage experiences a mutation in it's DNA. This mutated gene can then be transcribed and translated into a protein which may have an error that causes it to function improperly. This improper function can lead to cancer. If there is enough damage to the DNA, skin cells may also kill themselves; this is the reason for skin peeling after experiencing a sunburn.

Yeast is an excellent organism to use during experimentation because it has many alterable qualities and it is known to act as a model for mammalian biology. Yeast can be altered so that they cannot make specific types of repairs to their DNA. When yeast cannot repair DNA damage after UV exposure, they are prone to die. This allows one to observe the effects of being exposed or protected from UV light easily on yeast. Some properties that make yeast particularly suitable for biological studies include rapid growth, dispersed cells, and the ease of mutation (such as DNA-repair-efficiency like in this experiment).

In this experiment, DNA-deficient repair yeast colonies are exposed to UVB light under 4 different conditions. We chose to experiment with yeast colonies that are mutated in a way that they lack the enzyme necessary to repair DNA, the chemical in our cells that is essential for controlling proper growth and function. If the yeast were not mutated, it would require more UV exposure to observe its effects as the enzymes would allow the yeast to repair some damage done by the UV light. We chose to expose the yeast to artificial UV light because the amount of UV light transmitted by the sun that reaches the earth changes on any given day due to obstacles in the atmosphere such as clouds and the distance from the sun. Therefore, all our yeast received the same amount of UV exposure. In addition, we chose to expose the yeast under four different conditions in order to prove our hypothesis.

The four different conditions under which the DNA-repair-deficient yeast were exposed to had to do with the amount of sun protection factor (SPF) certain sunscreen had. In the first part of our experiment, the yeast colonies were exposed to Ultraviolet light with no protection. These were the yeast colonies that were compared with the protected colonies which did not undergo exposure to UV light; this comparison let us see the harmful effect that the UV light had on the yeast. In the next three conditions, the yeast was exposed to artificial UV light with the protection of either sunscreen with SPF 15, SPF 30, or SPF 45. This allowed us to see whether or not the UV light protection provided by sunscreen was effective in defending cells from damage.

The idea of sunscreen with different SPF's, or sun protection factors, is relatively modern. Many debate who the first chemist to begin experimenting with sun block and protection was. It is widely accepted that, Milton Blake, a South Australian chemist, experimented unsuccessfully with a sunburn cream in the early 1930's. Around the same time, another chemist, Eugene Schueller, found more success; he is often credited as the inventor of modern sunscreen.

The experimentation during the early 20th century allowed for sunscreen to be marketed for the first time in the late 1960's. However, this "sunscreen" was not as advanced as the sunscreen used today; the classification by SPF was not even introduced until 1972. Over a decade later, in the late 1980's, researchers finally determined that although UVB light causes most skin cancers, UVA rays also play a role in causing skin cancer. Therefore, in the early 1990's, sunscreens were improved to contain Ultraviolet-A blockers via an ingredient called Parsol 1789. Since then, many other chemicals have been found to protect the skin against UVA damage including: Avobenzone, Oxybenzone, Mexoryl, Zinc

Oxide, and Titanium dioxide.

As sunscreen continued to advance, chemists created sunscreens intended to be more protective, or, as seen on the bottle, have a higher sun protection factor (SPF). However, more recent studies show that this may not be true. According to James M. Spencer, MD, a Florida dermatologist and American Academy of Dermatology (AAD) spokesman, sunscreen with an SPF of 30 blocks about 97% of ultraviolet rays; while any protection above 30 remains at about 97% or 98%. This idea that increasing a sunscreen SPF beyond 30 will not help in preventing damage caused by UV light anymore than SPF 30 would has quickly spread and many beach-goers today use this idea when buying sunscreen. Therefore, we wanted to prove whether or not this was true.

When conducting this experiment, we had to account for variables other than the different SPFs of the four sunscreens which could cause our data to be inaccurate. First, it must be noted that when dealing with UV light, one usually deals with two different types (the third type, UVC, is of shorter wavelengths and cannot pass through the ozone layer, therefore it rarely reaches the earth's surface): UVA and UVB. This could affect our research because some sunscreens, regardless of their SPF, could provide more protection against UVA rays versus UVB rays or vise versa. Therefore, we made a point to conduct our experiment with all of the same brand and type of sunscreen. We chose to use broad-spectrum "CVS pharmacy ultra dry sheer lotion sunscreen" which provides both UVA and UVB protection. As many other scientists have pointed out, sunscreen protection is also dependent on the amount used. Therefore our results could be affected by differing amounts of sunscreen covering the yeast colonies. Thus, we made sure to apply the same amount of sunscreen to the different agar plates containing the yeast and to spread out the sunscreen evenly.

Scientists have many ideas about where they would like their experimenting with sunscreen and UV light to take them in the future. In general, scientists want to develop a pill that would protect the skin against harmful rays emitted by the sun capable of causing cancer. Scientists think that a pill would provide better protection against the sun because people often do not apply enough sunscreen and therefore do not even receive the minimum protection that the specified SPF is supposed to provide. Scientists are not only thinking about a pill, but also other antioxidants that could provide the accurate amount of protection needed to defend against UV rays. Experiments within the past decade have proved that, although most sunscreen provides excelled UVB protection, they do not protect the skin from UVA as much. Though it is true that UVB is generally thought of as more harmful, UVA light also contributes to skin cancer. Therefore, scientists would like to see UVA protection increased in all brands of sunscreen. The U.S. Food and Drug Administration (the FDA) would like to see an increase in UVA protection as well and in notifying consumers about the hazards of both UVB and UVA rays.

Throughout the past century, there has been great advancement in the protection that sunscreen provides against harmful Ultraviolet rays emitted from the sun. However, there is now debate over whether or not the labeling on sunscreens, or the SPFs, have really improved. It is now a widely accepted theory that sunscreen with SPF above 30 does not do a better job of preventing skin damage than SPF 30. Through this experiment we hope to prove whether or not this theory is correct.

Experimental Procedure:
Pouring the YED Plates

1. Loosen the cap of the YED (Yeast Extract Dextrose) agar bottle.
2. Heat the bottle in a microwave until the contents are completely melted.
   a. Stop the microwave to swirl the bottle every minute or so to keep the contents from boiling over.
3. Pour the melted YED agar into 13 petri dishes.
   a. Cover them immediately after pouring so the dishes remain sterile.

    b.   Protect your hands with heat-resistant gloves.

    c.   Pour just enough agar to cover the bottom of each plate.

    d.   Let the agar harden at room temperature overnight.

Streaking the Master Plate

1. Be sure your work area is clean and free of drafts.

2. Put on a pair of disposable gloves to keep the toothpicks sterile. Open one of the boxes of sterile toothpicks that came with the kit.

3. Touch the tip of a toothpick on an area where you can see yeast growing in the tube.

    a.   The yeast is shipped in a tube containing agar and nutrients (called the slant).

4. Glide the toothpick in a zigzag pattern across one of the YED plates that was poured previously.

5. Use a fresh toothpick (without yeast) to make another zigzag pattern through the first zigzag pattern. The idea is to get separate yeast cells that will grow into well-separated colonies.

6. Label the plate "Master Plate" with a permanent marker. Put the date on the plate.

7. Wrap the plate in aluminum foil to protect it from light.

8. Allow the yeast to grow for two days at room

temperature.

Labeling the Tubes and Making a Yeast Cell Suspension

1. Label four tubes, as follows:
    - 1. 1
    - 2. 1:10
    - 3. 1:100
    - 4. 1:1,000

2. After the two days have passed, put on a pair of disposable gloves and use a sterile toothpick to collect a mass of yeast from the master plate. The mass of yeast should be about 1 mm in diameter.

3. Smear the yeast inside the tube labeled "1," toward the bottom on the side of the tube.

4. Use a 5-mL bulb pipette to add 5 mL of sterile water to the tube with the yeast.

5. Shake the tube until the yeast are suspended.

Making Serial Dilutions

Use a new, different pipette for each of the following transfers:

1. Use the 3.5-mL sterile pipette to add 2.25 mL of sterile water into each of the tubes labeled 1:10, 1:100, 1:1,000 and 1:10,000.

2. Use a clean 1-mL pipette to transfer 0.250 mL of yeast from the tube labeled 1 to tube labeled 1:10.

3. Mix thoroughly.

4. Use a clean 1-mL pipette to transfer 0.250 mL

of yeast from the tube labeled 1:10 to tube labeled 1:100.

5. Mix thoroughly.

6. Use a clean 1-mL pipette to transfer 0.250 mL of yeast from the tube labeled 1:100 to tube labeled 1:1,000.

7. Mix thoroughly.

Labeling the Plates

1. Label the bottoms of twelve plates as follows:
a. SPF 8 normal
b. SPF 8 thicker
c. SPF 15 normal
d. SPF 15 thicker
e. SPF 30 normal
f. SPF 30 thicker
g. SPF 45 normal
h. SPF 45 thicker
i. SPF 100+ normal
j. SPF 100+ thicker
k. 1:1,000/control
l. 1:1,000/exposed

Spreading the Yeast onto Agar Plates

1. Remove the cover of each agar place and place 4-5 beads inside each plate. Put the cover on the plate

immediately after the beads are in.

2. Add 0.250 mL of yeast suspension from the tube labeled 1:1,000 to all the plates. It will be necessary to make another serial 1:1,000 serial dilution.

3. Shake the glass beads back and forth in order to spread the yeast cells.

4. Allow the plates 5-10 minutes to dry.

5. Remove beads from the plates by holding them vertically over a bowl and opening the cover so that the beads can drop out. The beads can be disposed of in a trash can.

Applying Sunscreen to the Plates

1. Cut out 12 portions of plastic wrap large enough to wrap around a petri dish.

2. Tare balance. Measure out both 1/16 tsp. and 1/2 tsp. (two different measurements) of each SPF 8, SPF 15, SPF 30, SPF 45 and SPF 100+ sunscreen.

3. Using gloved fingers, apply the sunscreen evenly to ten pre-cut pieces of plastic wrap. Each SPF should be applied to two pre-cut pieces of plastic wrap. There will be two pieces of plastic wrap without sunscreen. For each SPF, apply 1/16 tsp. to one plastic wrap and 1/3 tsp. to another.

4. Place the sunscreen-covered plastic wrap on their corresponding plates. (Ex. 1/16 tsp. SPF 30 goes on plate labeled, "SPF 30, normal")

5. Place the two sheets of plastic wrap without sunscreen on the two remaining plates. (These two plates should be labeled 1:1,000/control and 1:1,000/exposed).

Covering all Control Plates

1. Set aside the controlled petri dish.
2. Cover this plate with aluminum foil and set aside.

Exposing the Yeast to UVB Light
1. Expose the remaining 11 agar plates to a UV lamp.
2. Expose the plates for a total of one minute each.
3. After exposure, remove the plastic wrap from the plates and replace them with the lids.
4. Wrap the plates in aluminum foil to protect them from any light.
5. Let the plates sit at room temperature for 2 days.

Repeating the Procedure
1. Repeat the entire procedure to ensure that the results are reproducible.
2. To dispose of the yeast cultures, wrap the plates and tubes in aluminum foil and place of them in the regular trash.

Analyzing the Results
1. Unwrap the plates and count the number of colonies on each.

    a.   Each colony is formed from a single yeast cell.

    b.   Ideally, due to serial dilutions, one of the control plates should have about 100 colonies, as this number is small enough that you can count individual colonies, but large enough that you can get an accurate percentage of killed cells.

    c.   If the colonies are too close together to count, even at the 10,000-fold dilution, repeat the dilution series and add a 100,000-fold dilution.

2.   Graph the number of colonies for each plate. Put the number of colonies on the y-axis and the treatment and dilution on the x-axis.

3.   Calculate the percentage of cells killed by UV light. See Equation 1, below. Compare colony counts from plates with the same dilution.

    a.   Divide the number of colonies on the exposed plate by the number of colonies on the control plate.

    b.   Subtract this number from 1.

    c.   Multiply the resulting number by 100.

    d.   This yields the percentage of yeast killed by the sunlight.

Equation 1:

$$100 \times (1 - \text{colonies on exposed plate}/\text{colonies on control plate}) = \% \text{ killed}$$

(editors note: some charts removed)

Thinner (1/16 tsp.) Layer of Sunscreen Average % Yeast Colonies Killed

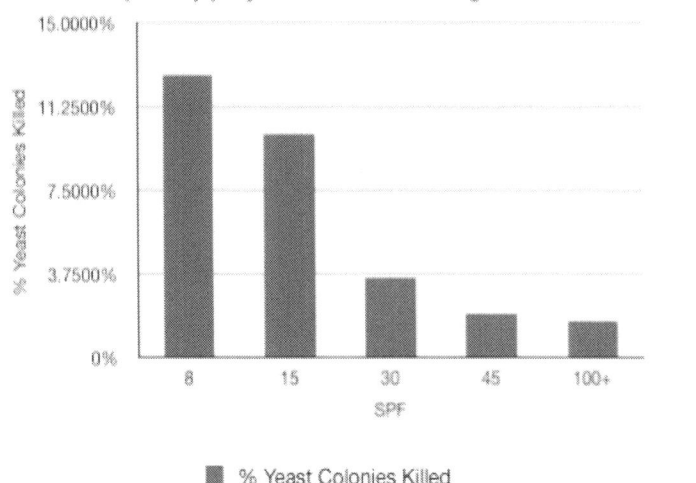

Discussion and Conclusion:

The hypothesis that yeast colonies would die after being

Thicker (1/2 tsp.) Layer of Sunscreen Average % Yeast Colonies Killed

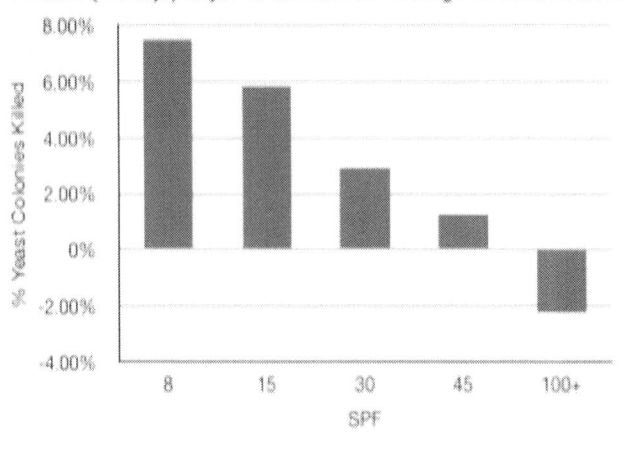

exposed to artificial UV light for one minute and be less damaged when sunscreen was used was supported by data found in our experiment. Our hypothesis that any SPF beyond 30 wouldn't affect how many yeast colonies are killed any more than SPF 30 was partially supported by data found in our experiment. Our hypothesis that the thickness of the sunscreen applied will not make a difference in defending yeast colonies from UV light was not supported by data found from our experiment.

The hypothesis that yeast colonies would die after being exposed to artificial UV light for one minute and be less damaged when sunscreen was used was supported in all our experiments and trials. We exposed an agar plate with DNA-repair deficient yeast on it to UV light for one minute and found that only two potential yeast colonies were present after two days of growth. We could not be sure if these were even yeast colonies as they were very small and could have been the result of slight contamination. Regardless, the serial dilutions were targeted so that around one-hundred yeast colonies would grow on each plate. Thus, nearly all the yeast were killed by direct exposure to UV light. When we applied sunscreen to plates with the same amount of yeast, our results showed that a significantly higher percentage of the yeast survived, thus supporting our hypothesis that sunscreen would minimize damage caused by UV light. We found that an average of 5.3% of the yeast were killed upon exposure to UV light with a thin layer of different SPFs of sunscreen, 98.06% of the yeast were killed upon direct exposure to UV light without any form of protection. Thus, a much higher percentage of yeast survived with sunscreen, demonstrating that sunscreen provides notable protection against the harmful rays of UV light.

Our hypothesis that any SPF beyond 30 wouldn't affect how many yeast colonies are killed any more than SPF 30 was somewhat supported by our experiment. Any SPF beyond 30 did not provide enough more protection to be of any significance. While 3.59% of the yeast were killed when exposed to UV light with a thin protection of SPF 30 sunscreen, 1.94% of the yeast died under the same amount of protection of SPF 45 and 1.65% died under the same amount of protection of SPF 100+. We predicted that it wouldn't be more effective at all, though it was actually more effective, but not enough so to be of any significance.

Our hypothesis that the thickness of the sunscreen applied will not make a difference in defending yeast colonies from UV light was not supported by our experiment, as we actually proved the opposite. We found that thicker layers of sunscreen provided much greater protection that thinner layers of sunscreen. While an average of 5.3% of the yeast were killed upon exposure to UV light with a thin layer of different SPFs of sunscreen whereas an average of 3.05% of the yeast were killed upon exposure to UV light with a thicker layer of the same SPFs of sunscreen. Thus, a thicker layer of sunscreen provided much more protection to the yeast cells than a thinner layer of sunscreen did.

We were not surprised to find that yeast colonies died upon exposure to UV light and sunscreen minimized the number of colonies that died. It is common knowledge that UV light is harmful to living organisms so we were expecting the yeast to undergo a large amount of damage from the light. Knowing that we were working with DNA-repair-deficient yeast which were more prone to die upon UV exposure, it was not surprising that direct exposure to UV light killed nearly all the yeast. It is also common knowledge that sunscreen minimizes damage from UV radiation, hence why so many people use it on sunny days. Therefore, we were not surprised to find that our data supported this hypothesis.

After studying sunscreens and sun protection factors (SPFs) further, it makes sense that our experiment concluded that any SPF greater than 30 provided more protection than SPF 30, though not significant. Our research of SPFs showed us that there is a fairly substantial difference in the amount of damage caused by UV light when dealing with smaller SPFs, such as those between SPF 1 and SPF 30. However, sunscreen with an SPF of 30 is already providing significant protection against UV light. Therefore, a higher SPF cannot provide much more protection than SPF 30 which already provides sufficient protection and the difference in damage done will be small.

Upon initially analyzing our results, we were confused as to why our hypothesis was not supported by our data the amount of damage caused by UV light is affected by how much sunscreen is applied. We were surprised to find that a thicker application of sunscreen provided significantly greater protection than a thinner application as we thought that it was really only the SPF level of a sunscreen that prevented UV damage. It is recommended that one applies sunscreen both liberally and frequently. After every two hours, swimming, towel-drying or heavily perspiring, sunscreen should be reapplied. One study cites that only twenty-five to fifty percent of people apply a sufficient amount of sunscreen. The repeated stress on the reapplication of sunscreen and the reapplication of sunscreen liberally led us to realize the importance of applying a sufficient amount of sunscreen in order to defend against UV rays. Therefore, it makes sense that our experiment concluded that a thicker application of sunscreen results in less damage caused by UV radiation.

There were a few variables during this experiment that could have gone wrong and affected our data. The biggest obstacle to overcome was contamination to the agar plates as we wanted to make sure that only yeast grew on the agar plates. Upon our first attempt at this experiment, we encountered problems with this as many of out plates became contaminated and we were unable to use them as a result. We think that this contamination was a result of taking the lids of the agar plates while pouring them, thus exposing them to the air and any contaminants which may have been in it. We learned from our mistakes in starting new trials as we were sure to take the lids off the plates only while pouring the Yeast Extract Dextrose (YED) and we placed the lid back on the plate immediately after pouring. It was frustrating to have to eliminate a whole trial due to contamination but we found that all our trials thereafter did not have any significant problems with contamination.

Another obstacle we had to overcome in trying this experiment was determining the time for which we should expose the yeast colonies to the UV light. We wanted to expose the yeast to UV light for long enough to cause damage to the yeast colonies and kill basically all of them if they were unprotected upon exposure to UV light yet not so long that it killed all of the yeast colonies that had sunscreen protection. We did many guess and check trials in which we exposed the yeast colonies to UV light and then grew the colonies for two days in order to find the perfect balance in which yeast colonies would be killed via direct exposure to UV light yet many of them would survive under the protection of sunscreen. We found that this period of time was one minute.

Another variable that may have affected our results was the serial dilutions that we created as it was the 1/1,000 dilution which contained the yeast that was spread onto all our agar plates and then exposed to UV light. These dilutions were targeted so that there would ideally be 100 yeast colonies to grow on each agar plate. However, this could have varied as the dilutions involved a set of many transfers from tube to tube and the yeast may not have been evenly dispersed in every tube, even after shaking. This accounts for many of our agar plates having different numbers of colonies even when they have the same SPF protecting them. We had to create a new set of serial dilutions for every trial, which may account for trial one having a lower number of yeast on all the plates overall in comparison to the other trials. For example, trial 1 (with a thin sunscreen layer) had an average of 93.8 colonies per plate while the third and second trials had higher averages of 98 and 98.8 colonies per plate. This means that the serial dilution used for trial one was likely less concentrated with yeast than the serial dilutions which were created and used for trials two and three. Our control plate was also found to have 103 yeast colonies, thus indicating that every plate would not have 100 colonies as the dilutions were not exactly perfect. One must account for a small percent error that is present when looking at our data.

From doing this experiment, we have learned a lot about the importance of using sunscreen and protection to prevent damage from exposure to UV light. It was interesting to see firsthand how organisms could be killed so quickly by UV light. We applied this knowledge to ourselves and our skin, which provided us with a greater realization of the damage UV light can cause to human beings and human skin. Our findings made it clear that sunscreen protection is vital in protecting against damage from UV rays. However, we also found that it is important to wear a high sunscreen SPF only to a certain extent; after increasing a sunscreen SPF beyond 30, one will receive little more protection. Rather, we realized the importance of applying enough sunscreen lotion and applying it liberally so that it does not come off, leaving a thin layer. By seeing how many more yeast died when exposed to UV light with a thinner layer of sunscreen as opposed to a thicker one, such as the fact that fewer yeast died when protected with a thick layer of SPF 8 than a thin layer of SPF 15, it was clear that applying sunscreen will not provide much protection if it is not applied correctly.

By seeing the harm caused by UV light, we can make better decisions about how to protect ourselves and our skin from harmful UV radiation emitted by the sun. Tanning booths, which emit large quantities of UV light, are clearly a bad choice as they can cause permanent damage to skin cells than can lead to skin cancer. Though it is well-known that tanning booths are a common cause of skin cancer, this project has taught us how to protect ourselves from UV radiation damage in other ways. From this experiment, we have concluded that SPF 30 is the smartest sunscreen to buy as it provides sufficient protection from harmful UV rays without costing one more to pay for a higher SPF sunscreen which will not provide much more protection. We have also concluded that the smartest way to use sunscreen is to apply of liberally and frequently as that will significantly reduce damage. After completing this experiment, we feel much more prepared to face sun exposure in a safe fashion that will not cause harm to our bodies.

Bibliography and Endnotes/ Footnotes

Bissonnette, R. "Update on Sunscreens: Protection Against Ultraviolet A (UVA) Radiation."Medscape Today. WebMD, 2008. Web. 11 Jan. 2011.

Elwood JM, Jopson J. Melanoma and sun exposure: an overview of published studies. Int J Cancer (1997). 73(2):198-203.

"Facts About Sunscreens." American Academy of Dermatology. American Academy of Dermatology, 2010. Web. 01 Feb. 2011.

Global Solar UV Index. World Health Organization. Pdf. 3 Feb. 2011.

"History of Sunscreen." Random History and Word Origins for the Curious Mind. 28 Apr. 2009. Web. 01 Mar. 2011.

"How Does Sunscreen Work?" Everyday Mysteries: Fun Science Facts from the Library of Congress. The Library of Congress, 23 Aug. 2010. Web. 23 Jan. 2011.

Louis, Catherine Saint. "Confused by SPF? Take a Number." The New York Times. 14 May 2009. Web. 14 Feb. 2011.

Neale, R, Williams, G, Green, A. Application patterns among participants randomized to daily sunscreen use in a skin cancer prevention trial.Arch Dermatol. 2002 Oct; 138, 1319-

1325.

O'Connor, Anahad. "The Claim: With Sunscreens, High SPF Ratings Are Best." New York Times. 7 Aug. 2007. Web. 22 Jan. 2011.

Office of Air and Radiation. "UV Radiation." SunWise Program. US Environmental Protection Agency, June 2010. Web. 01 Jan. 2011.

Parker-Rope, Tara. "Sunscreen Safety Called Into Question." New York Times. 22 July 2008. Web. 4 Dec. 2010.

Stenberg, Catharina, and Olle Larkö. "Sunscreen Application and Its Importance for the Sun Protection Factor, November 1985, Stenberg and Larkö 121 (11): 1400." Archives of Dermatology. AMA, Nov. 1985. Web.

Schaller, Bill, Teresa Herbert, and Rob Levy. "New Insight into Skin-tanning Process Suggests Novel Way of Preventing Skin Cancer - News Room." Children's Hospital Boston. Bill Schaller, Teresa Herbert or Rob Levy, 20 Sept. 2006. Web. 2 Feb. 2011

"Sunscreening Agents, Parsol 1789." 3D Chemistry. Aug. 2002. Web. 23 Feb. 2011.

"UVA Radiation: A Danger Outdoors and Indoors." The Skin Cancer Foundation. The Skin Cancer Foundation. Web. 01 Jan. 2011.

"What Is the History of Sunscreen?" Wiseminds. 3 June 2010. Web. 16 Jan. 2011.

Wood, T.d. "How to Choose and Use Sunscreen: Expert Advice from REI." REI. Recreational Equipment Inc., Mar. 2010. Web. 30 Jan. 2011.

The Efficacy of Thymol-enriched Fluconazole on *Aspergillus niger*
by Rachel Gottlieb

Abstract

Fungal infections are widely under publicized but can have serious consequences. Many fungi reproduce asexually with a similar generation time as bacteria and become drug resistant just as quickly. Antifungal drugs are intrinsically more antagonistic than antibiotics towards the human body due to the cellular similarities of fungi and animals. This project addresses both of these issues with the supplementation of fluconazole, a commonly-used systemic antifungal drug, with thymol, a concentrate of thyme oil. The two drugs were tested on *Aspergillus niger*, a opportunistic mold that can infect the lungs. The combination's efficacy was assessed by performing minimum inhibitory concentrations tests to compare the effectiveness of the enriched drug to the individual components. Preliminary tests show a possible synergy between the two drugs indicating that a significantly lower dosage of fluconazole could be equally effective. Future studies are needed to investigate potential side effects and anti-mutagenic properties to verify the drug combination could be used to increase rate of healing with minimal complications.

Introduction

The exploration of drugs and medicine has never been more

extensive. At the present, there are more drugs available for more conditions than ever before. However, as antibiotics improve and expand, the microbes they attempt to exterminate will only become more resistant to a wide spectrum of potent treatments. In fact, already with such outbreaks as MRSA and CRKP, many of the infected are faced with limited treatment options and a higher probability of fatality. Perhaps the ultimate symbol of this race with natural selection is the case of Penicillin, which thirty years ago was the cure-all and is now almost obsolete. It is this clear ephemeralness of prescription drugs that one is drawn to examine the use of natural remedies that have stood the test of time.

One of the most promising natural cures is the group of essential oils. In general, they have some very interesting properties. Not only are they cytotoxic against bacteria, fungus, and cancer but also inhibitory against viruses. In a study of twelve different essential oils, all extracts showed an inhibitory effect on the replication ability of HSV-1 when a 1% solution of the essential oil was incubated at 4°C for 24 hours[1]. In fact, lemongrass showed complete inhibition even at 0.1% concentration. Interestingly, research has also shown that essential oils can cause antimutagenic effects. One study found that "lavender oil exerted strong antimutagenic activity, reducing mutant colonies in the TA98 strain exposed to the direct mutagen 2-nitrofluorene" while also showing not mutagenic effects[2]. This could potentially mean that cancer cells could be treated and also prevented within a single treatment.

Plant oils are secondary metabolites that characteristically have high volatility and a chemical structure of C-10 H-16[3]. These remedies are already well entrenched in alternative medicine, because their organic structure causes decreased side effects. The general mode of action is through the fumes of the oil rather than physical contact[4]. At a cellular level, essential oils target the cell membrane, devastating both Gram positive and Gram negative models[5]. Experimentally, depletion of intracellular ATP and decreased pH were also observed.

Of all the essential oils, one of the most promising herbal extracts is thyme oil. In a comparison of ten essential oils on *P. acnes* and PC-3, A-549 and MCF-7 cancer cells, thyme oil showed a ring of inhibition of 40 +/- 1.2mm in comparison to the next most cytotoxic with rings of 33.5 +/- 1.5mm in cinnamon oil and 16.5 +/- 0.7mm in rose oil[6]. Researchers also found that thyme was fast acting with complete colony death after five minutes. As for the cancer cells, thyme oil showed the strongest cytotoxicity once again with inhibition concentration 50% (IC50) values of 0.010% (v/v), 0.011% (v/v) and 0.030% (v/v) on on PC-3, A549 and MCF-7 tumor cell lines respectively[6].

One particular case in which the cytotoxic capabilities of essential oils could be applied is in the treatment of fungi.

Though serious infections are significantly less prevalent, fungi are much more difficult to treat for several reasons. First, many fungi already possess drug resistance, and since most fungi, with the exception of basidiomycota, have an asexual cycle, their generation time is that of bacteria and therefore implies that fungi can develop resistance just as quickly. Secondly, the medicine used to treat cutaneous fungal infections is rather mild, but when systemic drugs must be used, severe side effects are more likely. This is due to the similarity of human and fungal cells as eukaryotes, and thus it is much more difficult to adequately address the infection without causing harm to the human tissue surrounding.

One example of this difficulty is the genus *Aspergillus*. Though this genus rarely infects, with a rate of incidence of 1-2 in 100,00 per year[7], in a study of English hospitals from the year 2002 to 2003, 74% of episodes of aspergillosis required hospital admission and 39% of incidents requiring emergency attention. The average stay of a patient infected with aspergillosis was 15.7 days while the median stay was 9 days, suggesting a wide variance in the length of the stay[8]. The genus itself is an opportunistic fungus that generally is prevalent in nature and content with consuming overripe food products. In fact, species such as *A. niger* are used economically and currently produce much of the citric acid used in industry. Its more pathogenic cousins tend to infect the lungs of immunosuppressed patients or those that work constantly around the spores. In addition, many secrete mycotoxins that are some of the most potent carcinogens on earth[9]. The typical symptoms of aspergillosis are coughing, wheezing, fever, shortness of breath, and sometimes bloody mucus. Invasive aspergillosis has the potential to infect and other part of the body[7].

Perhaps the cruelest part of this illness is that the treatment is almost as harmful as the infection itself. Fluconazole, a triazole also known as diflucan, is generally used in the case of systemic fungal infections. Its mode of action involved the inhibition of a protein in the membrane, resulting in a build in of incomplete proteins in the fungal cell in addition to inhibition of cell growth and sometimes cell death. Despite the drug's specificity, common side effects include headaches, dizziness, stomach pain, and diarrhea. Severe side effects can include swelling of the face and extremities, dark urine, seizures, and difficulty breathing. Despite this drug's effectiveness, it seemed that something must be done to ensure better patient health. The primary study surrounding this research project examined the supplementation of different essential oils with ketoconazole on species of *Trichopyton*. The results were striking as combinations of the antifungal prescription and the herbal extracts were anywhere from two times more effective to more than five times more effective than the components[10]. Such promising results ought to be applied to another genus that is in need of drug supplementation.

Materials and methods

The main goal of this experiment was to look at compare the efficacy of fluconazole and thymol, a concentration of thyme oil, to the two chemicals in combination. The efficacy of a given drug was be qualified through a minimum inhibitory concentration test. The MIC test will consist of a one fourth dilution of the given drug from 1 mg/mL to 1μg/mL.

Aspergillus niger was inoculated in 250 mL of yeast malt broth with a sterile magnetic stirring rod and several sterile (5-10) crystal beads. The inoculation was be constantly be stirred for 24 hours at room temperature in order to minimize the clumping of fungus. After 24 hours, the turbidity of the inoculation was be measured by the transmittance of a spectrometer with a set wavelength of 600nm[11]. Meanwhile, 38 test tubes were sterilized and set aside. Six of these tubes acted as a control dilution. They were filled with 0.9 mL of sterile isotonic saline solution. When the inoculate had been incubated, 0.1 mL was transferred into the first test tube of the control series with a 1/10th dilution. All six tubes were plated on Petri dishes with potato dextrose agar with four sections on one plate. Each quadrant had five 10 μL of one control test tube. The results of these control tubes were used to calculated how turbidity correlates to the CFUs/mL count.

Next, the fluconazole and thymol solutions were prepared. The fluconazole was prepared by combining 20 mL of sterile isotonic saline solution with 0.2 g of the fluconazole powder.

The thymol was prepared by combining 100 mL of sterile isotonic saline solution with .01g of crushed thymol crystals. 7.5 mL of inoculate was then added to twelve of the test tubes. In one set of six test tubes, 2.5 mL of the fluconazole was added and serially diluted by a factor of 1/4th. Each test tube was plated five times in increments of 10 μL on one quadrant of a potato dextrose Petri dish. Similarly, 2.5 mL of the thymol solution was added to the first test tube of the other series of six tube with the same dilution of 1/4th. Again, each test tube was plated five times in increments of 10 μL on one quadrant of a potato dextrose Petri dish.

Three test tubes were set apart. In one test tube, 8 mL of sterile isotonic saline solution was combined with 2 mL of fluconazole solution. In another test tube, 8 mL of sterile isotonic saline solution was combined with 2 mL of thymol solution. These two new solutions contributed 5 mL to the final test tube, resulting in a 50:50 split of fluconazole and thymol (1mg/mL of each). 7.5 mL of inoculate was added to another series of six sterile test tubes. 2.5 mL of the 50:50 split solution was added to the first tube of the series and serially diluted by a factor of 1/4th. Again, each test tube was plated five times in increments of 10 μL on one quadrant of a potato dextrose Petri dish.

Three test tubes were set apart once again. In one test tube, 7 mL of sterile isotonic saline solution and 1 mL of the original thymol solution were mixed. In another test tube, 7 mL of sterile isotonic saline solution and 1 mL of the original fluconazole solution were mixed. In the third tube, 2 mL of the first test tube and 8 mL of the second test tube were combined and mixed, creating a 20:80 split of thymol and fluconazole (1mg/mL of fluconazole and 0.25mg/mL of thymol). 7.5 mL of inoculate was added to another series of six sterile test tubes. 2.5 mL of the 20:80 split solution was added to the first tube of the series and serially diluted by a factor of 1/4th. Again, each test tube was plated five times in increments of 10 µL on one quadrant of a potato dextrose Petri dish.

All the above plate Petri dishes will be incubated for two days at room temperature. The control plates will be given an additional three days to grow in the absence of broth. The MIC of each solution will be recorded and a fractional examination will be done on them. This examination consists of dividing the MIC of the combination by the MIC of the fluconazole alone. This quotient will be added to the MIC of the combination divided by the MIC of the thymol alone. This sum represents the fractional inhibitory concentration index (FICI) of the MIC in combination. If the FICI is greater than 1, the combination is less effective than the components alone. If the FICI= 1, it is equally as effective as the components alone, and if the FICI is less than 1, the combination is more effective than the components alone.

Results

Due to inconsistent data, the results are best represented by a count of colonies per quadrant with an additional analysis of minimum inhibitory concentration test. First of all, Figure 1 offers an idea example of the colony count progression based on the dilution.

FIGURE 1

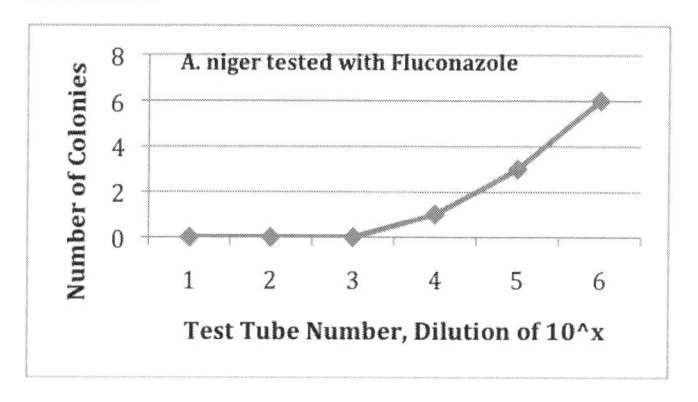

[Graph: A. niger tested with Fluconazole; Number of Colonies vs Test Tube Number, Dilution of 10^x]

Originally. the tests were done in 520nm, but inconsistent transmittance reading led to an increased wavelength to 600nm. Both are compiled upon each other, but eazch represents a different concentration of fungus (CFUs/mL).

FIGURE 2

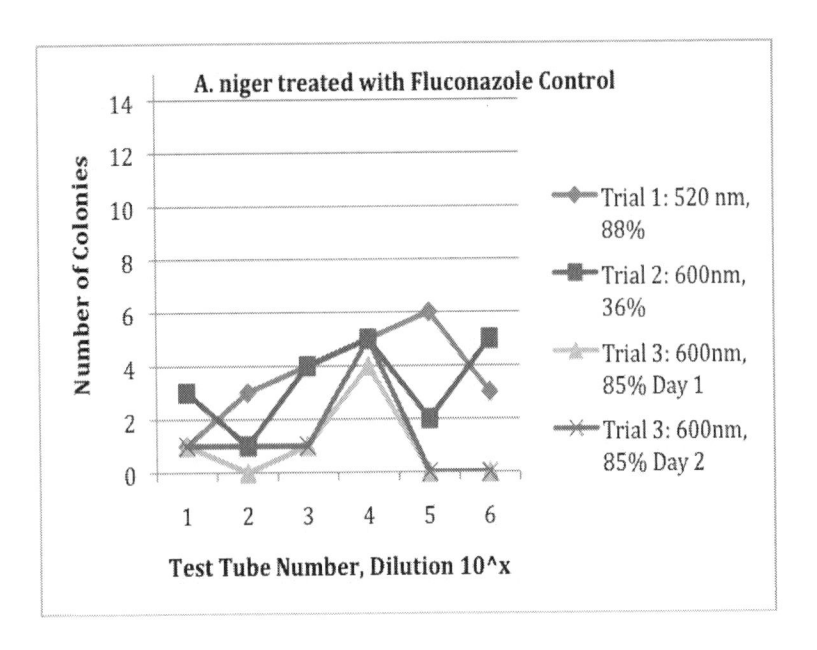

TABLE 1

Efficacy of Fluconazole Control on *A. niger* in Number of Colonies

Test Tube Number (Dilution= 10^x)	Trial 1 520nm, 88%	Trial 2 600nm, 36%	Trial 3 600nm, 85% Day 1	Trial 3 600 nm, 85% Day 2
1	1	3	1	1
2	3	1	0	1
3	4	4	1	1
4	5	5	4	5
5	6	2	0	0
6	3	5	0	0

FIGURE 3

A. niger treated with Thymol Control

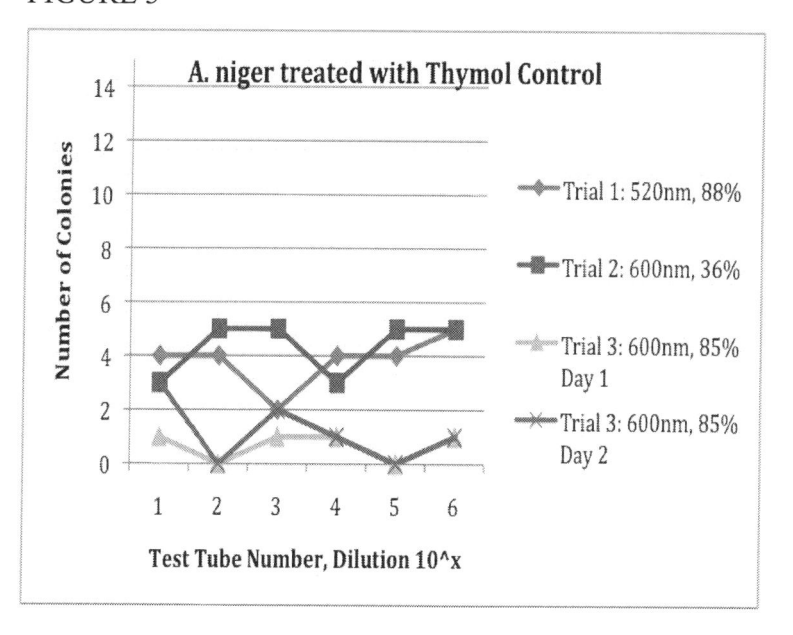

TABLE 2

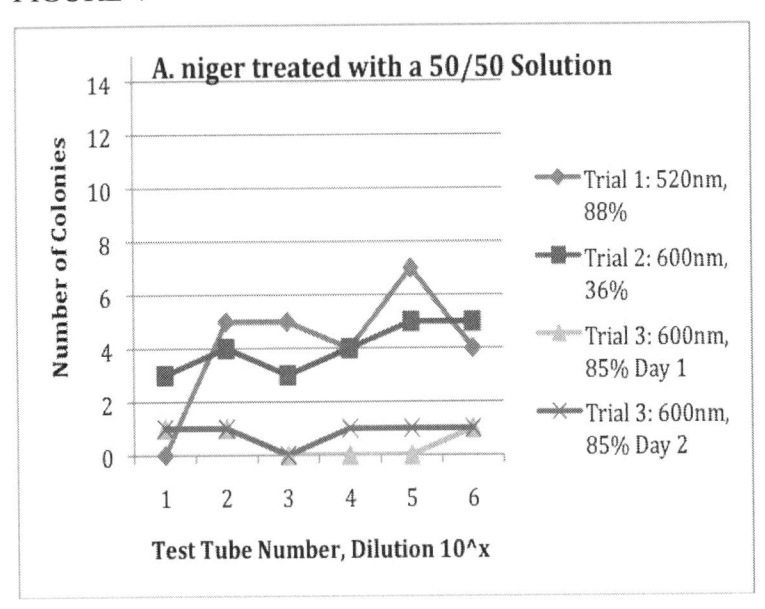

Efficacy of Thymol Control on *A. niger* in Number of Colonies

Test Tube Number (Dilution= 10^x)	Trial 1 520nm, 88%	Trial 2 600nm, 36%	Trial 3 600nm, 85% Day 1	Trial 3 600 nm, 85% Day 2
1	4	3	1	3
2	4	5	0	0
3	2	5	1	2
4	4	3	1	1
5	4	5	0	0
6	5	5	1	1

FIGURE 4

TABLE 3

Efficacy of 50/50 Split Solution on *A. niger* in Number of Colonies

Test Tube Number (Dilution= 10^x)	Trial 1 520nm, 88%	Trial 2 600nm, 36%	Trial 3 600nm, 85% Day 1	Trial 3 600 nm, 85% Day 2
1	0	3	1	1
2	5	4	1	1
3	5	3	0	0
4	4	4	0	1
5	7	5	0	1
6	4	5	1	1

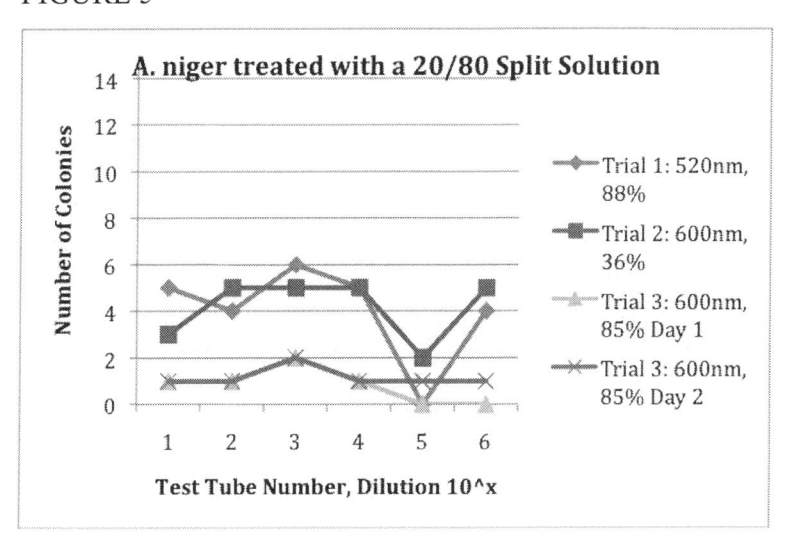

FIGURE 5

TABLE 4

Efficacy of 20/80 Split Solution on *A. niger* in Number of

Colonies

Test Tube Number (Dilution= 10^x)	Trial 1 520nm, 88%	Trial 2 600nm, 36%	Trial 3 600nm, 85% Day 1	Trial 3 600 nm, 85% Day 2
1	5	3	1	1
2	4	5	1	1
3	6	5	2	2
4	5	5	1	1
5	0	2	0	1
6	4	5	0	1

TABLE 5

Fractional Indices

Drug	MIC	FIC$_F$	FIC$_T$	FICI
Fluconazole	>.25mg/mL	N/A	N/A	N/A
Thymol	62.5µg/mL	N/A	N/A	N/A
50/50 Split	.25mg/mL	<1	4	<5
20/80 Split	>.25mg/mL	1	>4	>5

Because of the inconsistent data and MICs where inhibition was expected before that dilution, the estimates for the fraction indices were very conservative. Unless there are strong indications of excess fungus in a sample, the lowest MIC was taken. However, if the testing was redone and those high concentrations showed inhibition, the index could be much higher.

TABLE 6

Ideal Fraction Indices

Drug	MIC	FIC_F	FIC_T	FICI
Fluconazole	62.5 μg/mL	N/A	N/A	N/A
Thymol	62.5 μg/mL	N/A	N/A	N/A
50/50 Split	16.625 μg/mL	.25	.25	.5
20/80 Split	3.91 μg/mL	.125	.125	.25

Discussion and Conclusions

My original hypothesis and what past literature had suggested is that the combination would perform significantly better than the components individually. Even in the most ideal conditions, the combination was somewhat better at best. At worst, the combination is extremely ineffective comparatively. The difficulty with making conclusions in this case is the inconsistency of the data. For one, none of the tests were at the same wavelength and turbidity. Therefore, it is hard to know if this a fair representation of that fungal concentration. Even then, the MIC suggested by some of the data was simply illogical. The concept of a minimum inhibitory concentration suggests that there should be complete inhibition for all dilutions until the MIC. In my experiment, there would be no growth on some of the most diluted concentrations while also presenting growth on the high concentrations.

The major error that I suspect skewed the data was the almost impossibility of a representative sample. In order my method to work, the inoculated broth had to have complete homogeneity. It was my hope that the fungus would behave like the *Trichophyton* genus, but instead of a general opaqueness, the fungus would clump into small but very visible groups (note that this was when the broth was constantly stirred and with crystal beads to break up clumps). Therefore, when I plated, either a large clump would be absorbed or almost on fungus at all. Even from test tube to test tube, the concentration of fungus would vary as without constant mixing, the fungus sinks to the bottom. My general conclusion is that the method for the MIC test was not well suited for *A. niger*.

Perhaps the most disappointing aspect of this experiment was fluconazole's general ineffectiveness. What I suspect is that this triazole works first as an inhibitor, with the first day presenting no growth consistently, and later on kills the cell with build up. However, only a day's worth of fluconazole was applied so perhaps it was all consumed by the second day when growth was recorded.

Thus, in the case of applications, it is difficult whether synergy was proved or even suggested. The best action going forward would be to find a test that better accommodates the growth patterns of *A. niger*. In addition, perhaps a different systemic antifungal drug that primarily kills instead of inhibits should be used. Though this experiment was not very successful, I feel confident that the synergy between antifungal drugs and thymol can and will be proven with more testing. In addition to future exploration of this pairing, further examination of thymol's side effects would be necessary if it were to be supplemented as a side effects reducer. One must verify that side effects are decreased instead of different side effects being added on. Finally, more potent species of the *Aspergillus* species should be examined to see how effective the drug combination is on those. As the study "Antifungal effects of herbal essential oils alone and in combination with ketoconazole against Trichophyton spp..", showed, there was a significant variance within the *Trichophyton* genus, with FICIs from 0.56 and 0.18 with the same drug combination and concentration.

Overall, I am generally optimistic that with a better procedure and more consideration for the mode of action of fluconazole, synergy between thymol and fluconazole can be shown.

References

[1] Minami, M, M Kita, T Nakaya, T Yamamoto, and H Kuriyama. "The inhibitory effect of
essential oils on herpes simplex virus type-1 replication in vitro." *Microbiological*

Immunology 47.9 (2003): 681-684. Web. 11 Oct 2010.

[2] Evandri, M.G., L Battinelli, C Daniele, S Mastrangelo, and P Bolle. "The antimutagenic activity of Lavandula angustifolia (lavender) essential oil in the bacterial reverse mutation assay." *Food and Chemical Toxicology* 43. (2005): 1381-1387. Web. 11 Oct 2010.

[3] Upadhyay, R.K. "ESSENTIAL OILS: ANTI-MICROBIAL, ANTIHELMINTHIC, ANTIVIRAL, ANTICANCER AND ANTI-INSECT PROPERTIES." *Journal of Applied Bioscience* 36.1 (2010): 1-22. Web. 11 Oct 2010.

[4] George, D.R., T.J. Smith, R.S. Shiel, O.A.E. Sparagano, and J.H. Guy. "Mode of action and variability in efficacy of plant essential oils showing toxicity against the poultry red mite, Dermanyssus gallinae." *Veterinary Parasitology* 161.3-4 (2009): 276-282. Web. 11 Oct 2010.

[5] Oussalah, M, S Caillet, and M Lacroix. "Mechanism of action of Spanish oregano, Chinese cinnamon, and savory essential oils against cell membranes and walls of Escherichia coli O157:H7 and Listeria monocytogenes.." *Journal of Food Protection* 69.5 (2006): 1046-1055. Web. 11 Oct 2010.

[6] Yuangang, Zu, Yu Huimin, Liang Lu, Fu Yujie, and Efferth Thomas. "Activities of Ten Essential Oils towards Propionibacterium acnes and PC-3, A-

549 and MCF-7 Cancer Cells." *Molecules* 15. (2010): 3200-3210. Web. 11 Oct 2010.

[7]"Aspergillosis." *Center for Disease Control and Prevention.* CDC, 20 Jul 2010. Web. 13 Mar 2011.

[8]Sneddon, Andrew. England. *Hospital Episode Statistics.* London: Government Statistical Service, 2003. Web. 13 Mar 2011.

[9] Bennett, Joan W. "An Overview of the Genus Aspergillus."*Aspergillus Molecular Biology and Genomics*. 'Ed'. Masayuki Machida. Ibarakl, Japan: National Institute of Advanced Industrial Science and Technology, 2010. Ditial.

[10] Shin, S, and S Lim. "Antifungal effects of herbal essential oils alone and in combination with ketoconazole against Trichophyton spp.." *Journal of Applied Microbiology* 97.6 (2004): 1289-1296. Web. 11 Oct 2010.

[11] Santos, D.A., M.E.S. Barros, and J.S. Hamdan. "Establishing a Method of Inoculum Preparation for Susceptibility Testing of Trichophyton rubrum and Trichophyton mentagrophytes."*Journal of Clinical Microbiology* 44.1 (2006): 98-101. Web. 1 Nov 2010.

A Tasty Treat, or Unrecognized Reinforcer?

By Evan Grandfield

"I began thinking about possible science fair options the summer before the task was assigned. After reading some works by B.F. Skinner, I realized I wanted to test positive reinforcers, against aversive training, to see which was more effective in altering rats' behavior. I quickly purchased the rats and began testing. Although mostly satisfied with my project, I later realized the need for more test subjects. I advise anyone thinking about using rats as subjects to find a psychology lab and gain permission to use its facilities. At competition at the state level, having a professional test environment makes all the difference."

Purpose of Research

The purpose of the research is to determine whether positive reinforcement or punishment is more effective in altering behavior. Especially important is whether or not punishment in the form of bitter food has an effect on the behavior of the rat, as bitter food is not a very potent deterrent of behavior.

Hypothesis

Punishment will have a greater effect on the behavior of lab rats than positive reinforcement will.

Concise Description of Experiment

There will be two rats, one exposed to a punishment and the other exposed to a positive reinforcement. Alternatives for rats may include fish or invertebrates, but these species do not possess the same mental capacity as do rats. The genus and species of the rats used is *Rattus* norvegicus.

1. The rat (Subject A) exposed to punishment is placed in a separate cage with two bowls, one higher than the other.
2. By reaching for the lower bowl, the rat gains access to a food pellet that is coated with a bitter solution (Bite-It) that the rat finds distasteful.
3. In order to ascertain the pellet more suitable for consumption, the rat has to access the bowl that is higher and more difficult to reach. The distasteful food reinforces it to reach for the higher bowl.
4. The rat (Subject B) exposed to positive reinforcement is placed in a separate cage with the same bowls, but both contain the uncoated pellets.
5. The food tastes the same from both bowls, but after so many pellets are taken (six pellets) the lower bowl will be emptied. Consequently, the rat is positively reinforced to go to the higher bowl, where there are ten pellets.
6. After three conditioning attempts in their respective environments, the rats are transferred individually into a cage with the same two bowls, but with each containing uncoated pellets with the same supply.
7. The researcher then records how the two conditioned rats

responded to their conditioning and later the testing environment. The researcher must analyze the results to find which rat reaches for the higher bowl more times before reaching for the lower bowl. These steps are repeated two more times.

Expected Outcome

The expected outcome is that the rat subjected to punishment will have its behavior modified more than the rat subjected to positive reinforcement. This means the rat subjected to punishment will respond to its testing environment by taking food pellets from the higher bowl with a greater frequency than it will take food pellets from the lower bowl.

Significant Results

As the conditioning attempts progressed, it was noted how the Subject A's (punishment) behavior was altered in the conditioning environment. As the rat experienced the bitter taste of the lower-placed food pellets, it ceased to take those pellets and only reached for the higher-placed pellets. Subject B's behavior (positive reinforcement) in the conditioning environment was not altered by the positive reinforcement in the form of the large supply of higher-placed food pellets. Every attempt he ate all the available food pellets, with little preference of one of the bowl's over the other. Often, it ate all of the lower pellets first, which may have shown positive reinforcement's deficiency, but in conditioning attempts seven, eight, and nine, Subject B ate many of the higher pellets before proceeding to eat the lower pellets.

In the first and third test attempts, Subject A responded by eating all the higher-placed pellets and then hesitating for a few minutes before sniffing the lower-placed pellets. He then ate every lower-placed pellet. In the second test attempt however, he switched between bowls before he was done with either one of them.

Subject B responded to the first test attempt by eating all of the lower-placed pellets before eating all of the higher-placed pellets. In the second test attempt, he ate all of the higher-placed pellets before proceeding to the lower-placed pellets. He also did this in the third test attempt.

Introduction

Rationale

Humans constantly create environments so that they can get the most out of themselves and those working with and below them. Teachers try to create classroom environments that foster growth in pupils, and supervisors attempt to control working conditions to increase profit. Despite the effort put forth to create successful environments, many disagree on whether positive reinforcement or punishment has a greater effect on behavior. In order to make more practical working and learning environments, the validity of positive reinforcement and punishment must be examined. Therefore, this experiment has been conducted to definitively reveal which has a greater influence on the psyche and behavior of individuals.

Background

In this science fair experiment, *A Tasty Treat, or Unrecognized Reinforcer?*, the behavior of two independently-conditioned lab rats is monitored and recorded to answer the question of whether punishment or positive reinforcement is more effective in altering behavior. In this experiment, two male rats are subjected to different environments, one exposed to punishment and the other exposed to positive reinforcement.

The rats used in this experiment are of the species *Rattus norvegicus*. The *Rattus norvegicus* makes up the majority of pet rats and lab rats, and also inhabits every continent excluding Antarctica ("Rat Systematics," n.d.). This species is a member of the mouse family and its members can live for about four years. The tail and ears of the rats are bald, and their fur is usually brownish, but domesticated individuals can be albino. The commonness of the *Rattus Norvegicus* makes it an ideal organism to use in psychological studies, in which individual experiments are often repeated by many different researchers.

This experiment has its roots in behaviorism, which is a type of psychology that is based on the principle that conditioning produces all behavior. The two types of conditioning are classical and operant (instrumental) (Sherry, n.d.a). Classical conditioning is best explained by the example of a dog's response to the smell of food. When a dog smells the odor of food, the unconditioned stimulus, it responds by salivation, the unconditioned response. Stimuli are defined as things that occur in the environment of an individual and influence its behavior (Barker, Kreider, Peissig, Sokoloff, & Stansfield, n.d). If, for instance, a bell is wrung when the food is presented, the bell acts as a conditioned stimulus, whereas before it would have been a neutral stimulus. After many repetitions of the food being presented with the ringing of the bell, the bell itself may elicit the salivation from dog, even though it is the stimulus of the food's odor that originally made the dog salivate. The salivation of the dog in response to the bell's ringing is called the conditioned stimulus (Sherry, n.d.b). The work of Russian physiologist Ivan Pavlov actually mirrored the example used; however he replaced the bell for a metronome as his conditioned stimulus (Sherry, n.d.c)

However, this science fair experiment utilizes operant conditioning rather than classical conditioning. Operant conditioning is based on the principle of administering reinforcers to increase a behavior and administering punishments to decrease a behavior. In this experiment, positive reinforcement and positive punishment will be used to alter behavior, but the terms negative reinforcement and negative punishment are sometimes misused and do not apply to this experiment. Positive reinforcers are defined as rewards given after a behavior, in order to make that behavior occur more often. A good example would be a teacher giving a student extra credit for washing the blackboard. By giving the student extra credit, the teacher is "reinforcing" the student to wash the blackboard. Negative reinforcement is defined as the elimination of an aversive thing or event in order to make a certain behavior occur more often (Sherry, n.d.b) An example is found in the act of beating a horse until it moves. By moving, the beating ceases and the horse is reinforced to move, thereby preventing pain. The lashing is the aversive contingency, or the unpleasant condition for this horse. Positive punishment is the introduction of something aversive in order to decrease the occurrence of a behavior. An example is found in administering a shock to someone every time he says a certain word. In order to avoid the pain of the shock, the person abstains from using that certain word. Negative punishment is defined as eliminating a desired or coveted thing or event in order to decrease the occurrence of a behavior (Barker, Kreider, Peissig, Sokoloff, & Stansfield, n.d). An example of this is taking away a student's athletic privileges when he fails a course. In order to avoid the aversive consequence of having his athletic privileges taken away, the student does not fail any more of his courses. In this science fair experiment, one rat is positively reinforced by extra food, so that he should reach for the higher bowl. The other rat is positively punished by the addition of the bitter-tasting food; therefore he should reach for the higher bowl to get normal food.

This concept of operant conditioning was first developed by American psychologist B.F. Skinner, who invented the operant-conditioning chamber (Skinner box), so that he could experiment with rats to find how behavior could be increased or decreased. He believed behavior was the direct product of external experience and not of internal thoughts and emotions, which were not autonomously formulated, but caused by external stimuli (B.F. Skinner, n.d). The experiment performed in this science fair falls in context to the experiment of Luke S. Watson and Reed Lawson, *Learning in the Rat (Rattus norvegicus) under Positive Vs. Negative Reinforcement with Incentive Conditions Controlled.* In this experiment, two groups of rats received an auditory stimulus that was followed by the presentation of a food pellet. The presentation of the food pellet was dependent upon their correct response to the sound by depressing the lever. For one group it was a desirable pellet, but for the other group it was an undesirable pellet (respectively, they were labeled high and low reward groups). The third group received the same auditory stimulus, but had to depress the lever in a five second time period in order to avoid an electric shock. In two groups, positive reinforcement was used to have the rats depress the lever, but in the third group, negative reinforcement was used to have the rats depress the lever in order to avoid the shock. While this experiment proved that positive reinforcement was more effective than negative reinforcement, this science fair experiment is predicted to find that punishment will be more effective than positive reinforcement. While the experiment of Watson and Lawson used electric shock as an aversive consequence, this experiment will use a bitter food pellet (Lawson & Watson, 1963, pp. 87-91). The substitution of the shock for the bitter food helps this science fair experiment model the punishments of everyday life, which usually are less aversive than electric shock.

This project has had its inspiration in the work of B.F. Skinner, whose book, *Beyond Freedom and Dignity*, provided the idea for testing the effectiveness of positive reinforcement and punishment. In this famous treatise, the renowned psychologist champions what was later called "radical behaviorism"; his belief in this principle led to him to say that all human emotions, intentions, and goals are simply products of reinforcers and punishments. He believed that the "autonomous man" governed by free will did not exist, and was rather a product of the environment around him (Doyle, n.d.). These avant-garde convictions aroused an interest in psychology in the researcher and encouraged him to construct his experiment.

Question and Hypothesis

Will positive reinforcement or punishment have a greater effect on the behavior of the lab rats? The hypothesis is that punishment will have a greater effect on the behavior of lab rats than positive reinforcement will.

Materials and Methods

Materials

The materials used by the researcher are:

2 Petco rat water bottles

2 (14''X8''X9'') small animal cages (including rat wheels)

1 stopwatch function on the researcher's Samsung Gravity ® 3 cell phone

2 packages of Petco brand Munchie chew blocks (24 blocks per package)

2 packages of Petco brand paper bedding

2 packages of Kaytee® Forti-Diet® small animal food (5lb bags)

2 packages of Kaytee® Fiesta® blueberry yogurt-covered treats

4 small Petco brand bird bowls (can attach to bars on cage)

2 3'' diameter glass food dishes

2 rats (*Rattus norvegicus*)

1 carving knife

1 bottle of Bite-It® nail-biting deterrent

1 pair of plastic gloves (for handling the rats)

Experimental Setup and Directions

Once the rats have been purchased, they are placed in identical 14''X8''X9'' cages with rat wheels placed in the back, left side of the cage. They are fed a bowlful of Kaytee® Forti-Diet® small animal food daily and there paper blend bedding is replaced weekly. Their water bottles are to be placed on the left wall of their cages, and replenished as necessary. They are left in their cages for approximately two weeks before they are subjected to conditioning, as they need to be well-adjusted to their new cages.

The researcher must conduct three conditioning attempts with the rats before conducting one test attempt. This cycle of conditioning and testing is repeated three times in total so that the data can be considered reliable. There must be no more than one attempt per day, as more attempts would endanger the rats' digestive health.

Directions (conditioning attempt):

1. To begin each conditioning attempt, the researcher must cut 14 Kaytee® Fiesta® blueberry yogurt-covered treats into halves using a carving knife.
2. The researcher must then coat six of the halves with Bite-It® nail-biting deterrent. The halves should be coated liberally, but not so that they are dripping with the solution.

3. The researcher must take the 28 halves and distribute them among four plastic bowls. Eighteen of these are divided evenly into groups of six, and distributed among three plastic bowls (the coated pellets should all be in the same bowl). The researcher must put the remaining ten halves into a fourth plastic bowl.
4. The bowl containing the six coated pellets is put into Subject A's cage by clicking the metal pieces onto the second lowest metal bar. The bowl should be positioned so that it is on the left side of the cage. Another bowl containing six uncoated pellets in clicked onto the fourth highest bar in Subject A's cage. It should be positioned far to the right of the lower bowl.
5. The researcher must then put one bowl containing six uncoated pellets and one bowl containing ten pellets into

Subject B's cage. The bowl containing the six uncoated pellets is put into the cage according to the method prescribed to Subject A's lower bowl. The bowl containing the ten pellets is put into Subject B's cage by the method prescribed to Subject A's higher bowl.

6. The researcher must then remove the glass food bowls from each cage and turn off the lights in the room he is working in. The researcher proceeds to observe the rats and make notes regarding their responses to the two bowls. He must record the amount of time it takes for them to eat the pellets, concluding the timing after each rat has ceased to take from the bowls.

7. Once the conditioning attempt is concluded, the researcher must remove the plastic bowls from the cages and return the glass bowls, filling them with the normal food. Post-attempt observations of the attempt should be recorded so as to summarize the experiment.

Directions (test attempt):

1. To begin each of the three test attempts, the researcher must cut 12 Kaytee® Fiesta® blueberry yogurt-covered treats into halves using a carving knife.

2. The researcher then distributes the 24 halves evenly into four plastic bowls.

3. In each cage, one bowl is clipped to the second lowest bar

and another is clipped to the fourth highest bar. The lower bowl is positioned to the left and the higher bowl is positioned to the right.

4. The researcher must then remove the glass food bowls from each cage and turn off the lights in the room he is working in. The researcher proceeds to observe the rats and make notes regarding their responses to the two bowls. He must record the amount of time it takes for them to eat the pellets, concluding the timing after each rat has ceased to take from the bowls.

5. Once the conditioning attempt is concluded, the researcher must remove the plastic bowls from the cages and return the glass bowls, filling them with the normal food. Post-attempt observations of the attempt should be recorded so as to summarize the experiment.

Results

Written Description

In the conditioning attempts, Subject A ate approximately 26% of the lower-placed pellets, but ate 100% of the higher-placed pellets. Subject B ate 100% of both the lower- and higher-placed pellets. During the conditioning attempts, Subject A did not eat any of the lower-placed pellets during 56% of the attempts and only ate all of them during 22% of the attempts. Subject A took an average five minutes to begin eating the higher-placed pellets and an average of 12 minutes and 12 seconds to finish the higher-placed pellets. Subject B took an average of two minutes and 12 seconds to begin eating the lower-placed pellets and took an average of 11 minutes and 42 seconds to finish the lower-placed pellets. Subject B also took an average of three minutes and 54 seconds to start eating the higher-placed pellets and an average of 14 minutes and 24 seconds to finish the higher-placed pellets.

During the conditioning attempts, Subject B ate 385% of the amount of lower-placed pellets that Subject A ate. Subject A took one minute and six seconds longer to begin eating the higher-placed pellets than did Subject B. Subject A also took two minutes and 12 seconds more to finish the higher-placed pellets than did Subject B.

In the testing attempts, both Subject A and Subject B ate 100% of the presented food pellets. However, Subject A took an average of five minutes to begin eating the lower-placed pellets and an average of 16 minutes and 42 seconds to finish them. Subject A also took an average of one minute and 42 seconds to begin eating the higher-placed pellets and an average of seven minutes to finish them. Subject B took an average of zero minutes to begin eating the lower-placed pellets and an average of six minutes and 42 seconds to finish them. Subject B also took an average of zero minutes to begin eating the higher-placed pellets and an average of six minutes and 42 seconds to finish them.

This means that during the test attempts, it took Subject A five more minutes than Subject B to start eating the lower-placed pellets and ten more minutes to finish them. It took Subject A one minute and 42 seconds more to start eating the higher-placed pellets than it took Subject B to begin eating them. Subject A finished the higher-placed pellets only 18 seconds later than did Subject B. This shows that on average, it took Subject A longer to respond, but the delay was less for the higher-pellets than the lower-pellets.

Tables and Graphs

Graphs are in chronological order of conditioning with testing
Subject A=punishment
Subject B=positive reinforcement
L=low pellets
H=high pellets

Conditioning Attempt #1-Figure 1

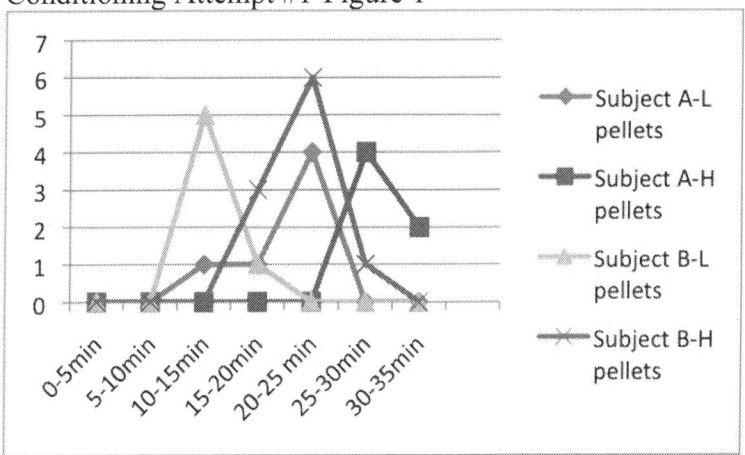

Conditioning Attempt #2-Figure 2

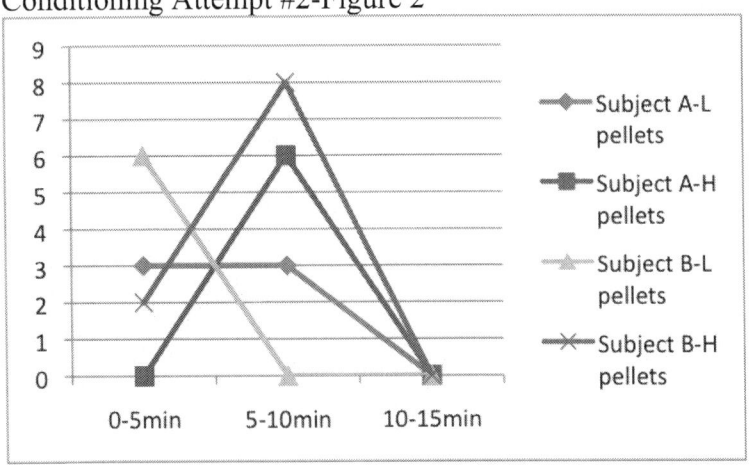

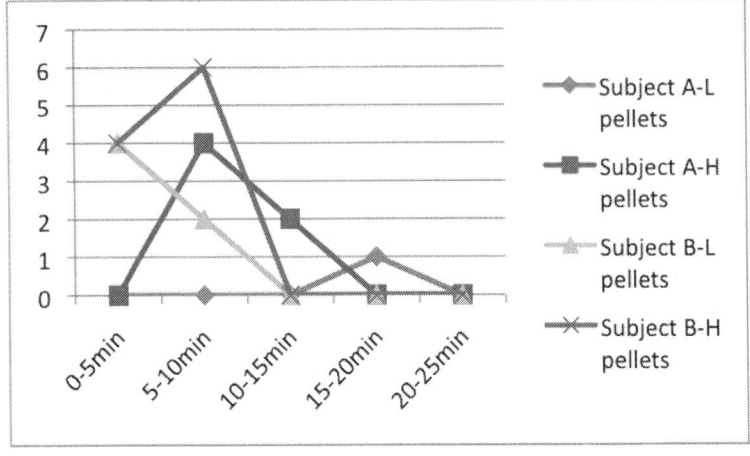

Conditioning Attempt #3-Figure 3

Test Attempt #1-Figure 4

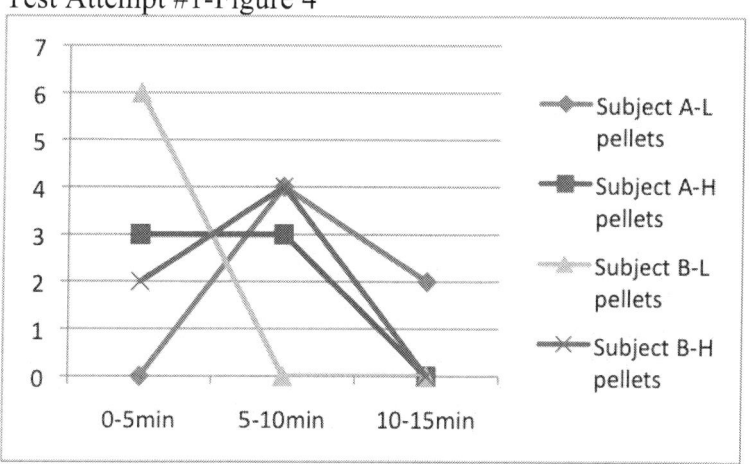

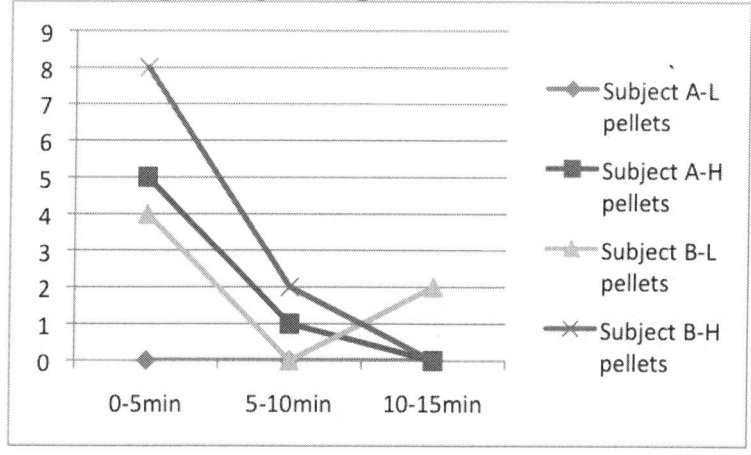

Conditioning Attempt #4-Figure 5

Conditioning Attempt #5-Figure 6

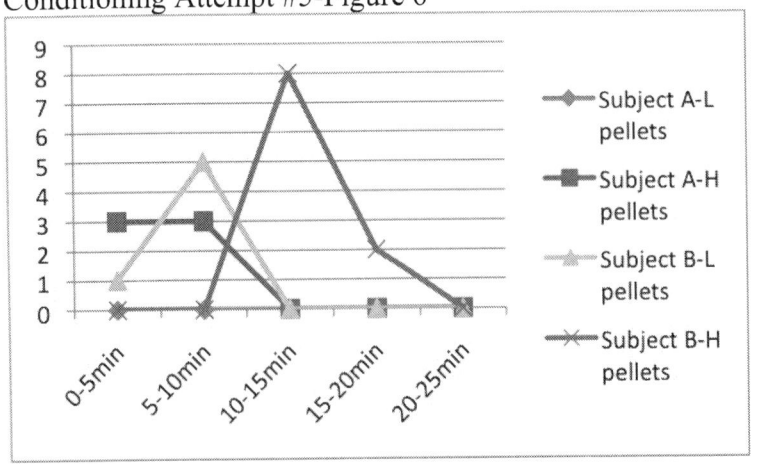

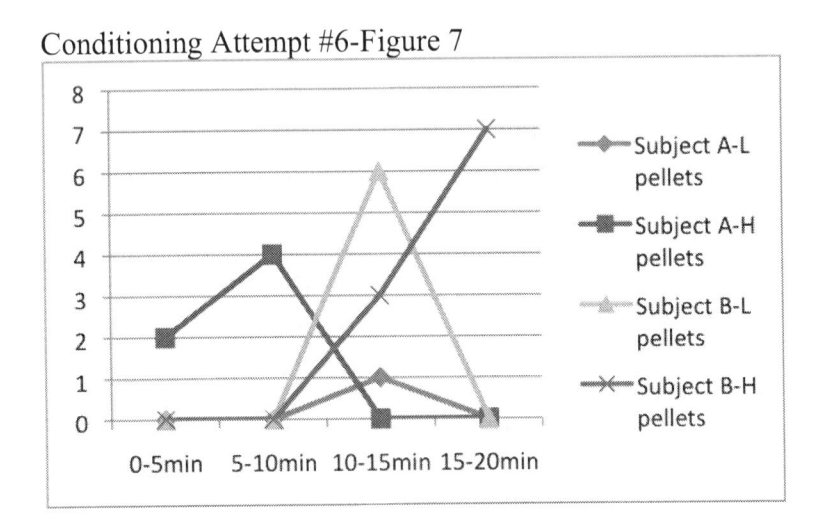

Conditioning Attempt #6-Figure 7

Test Attempt #2-Figure 8

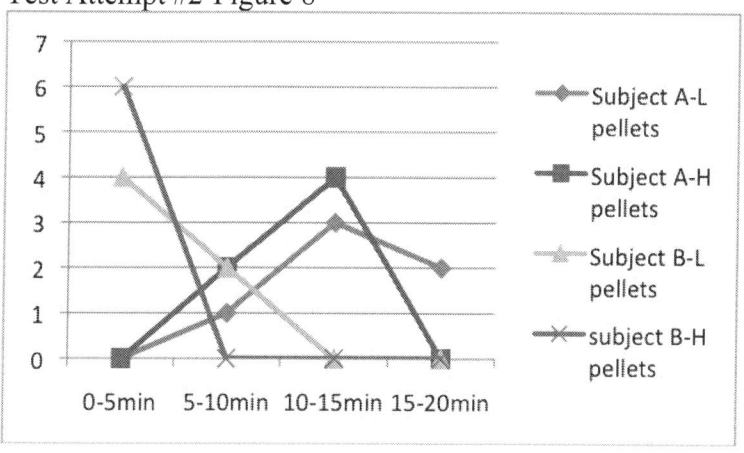

Conditioning Attempt #7-Figure 9

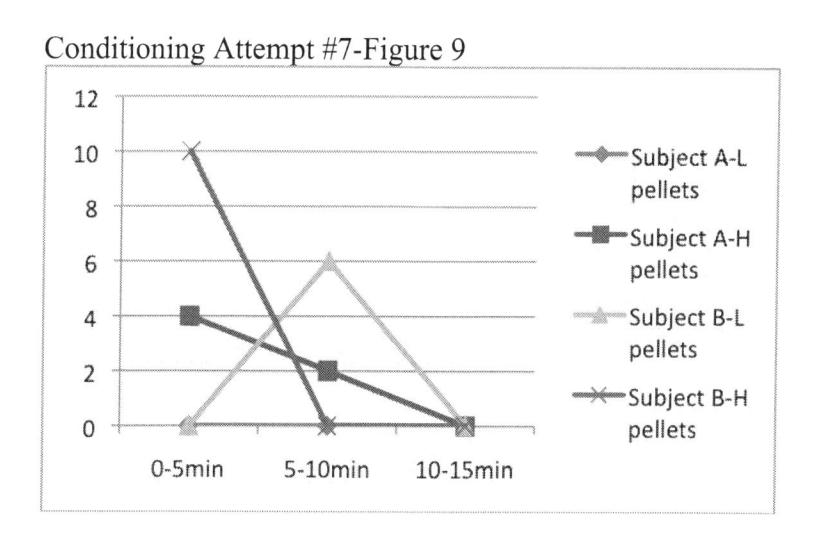

Conditioning Attempt #8-Figure 10

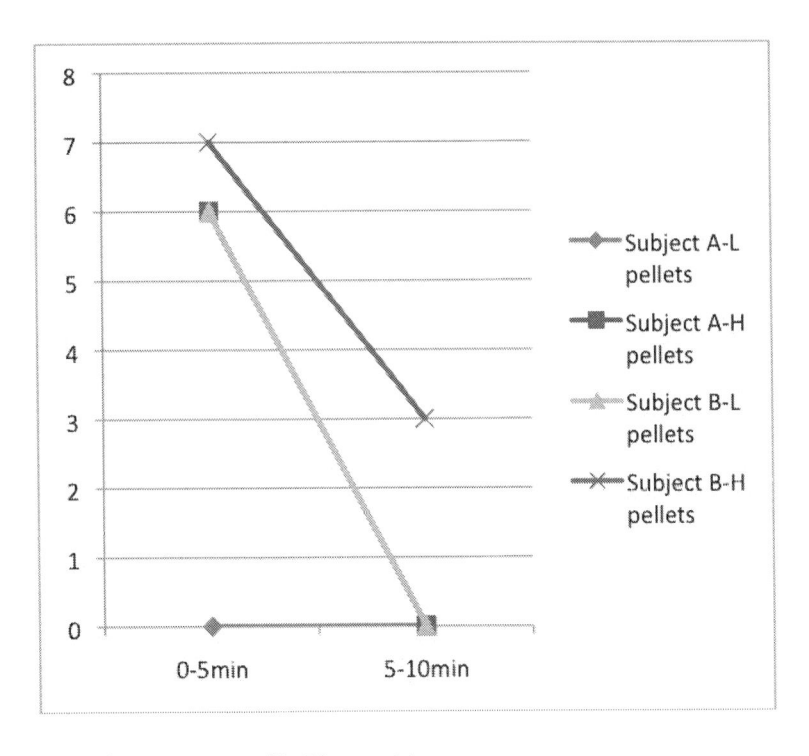

Conditioning Attempt #9-Figure 11

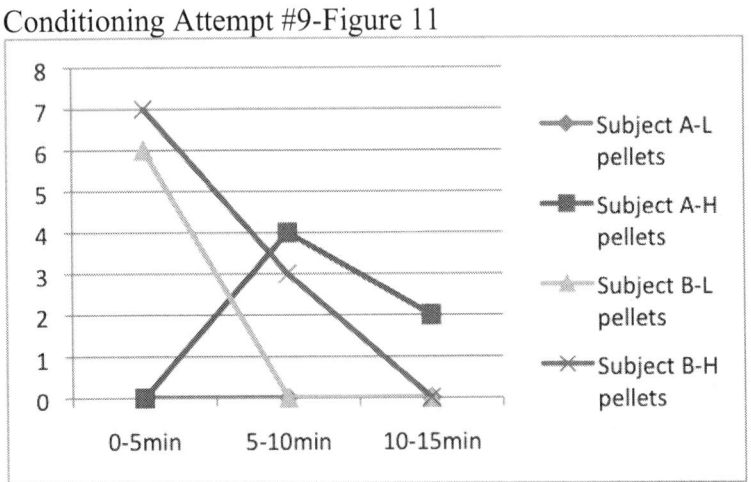

Test Attempt #3-Figure 12

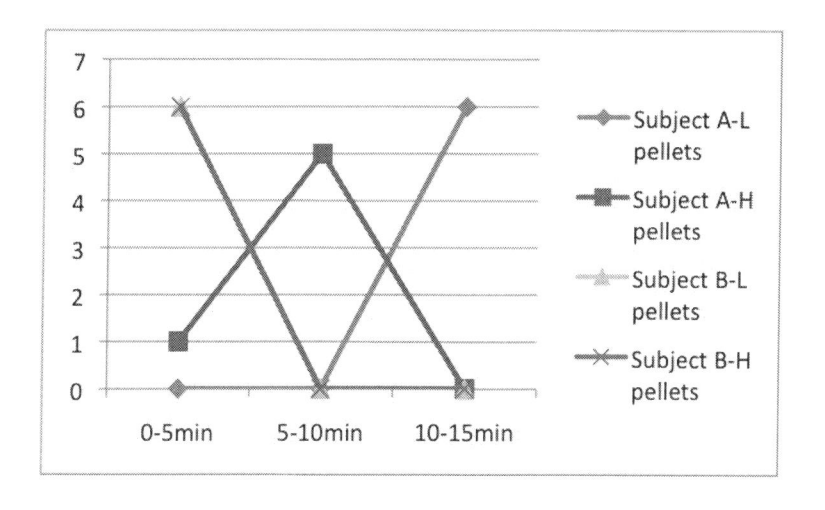

Conclusion

Results Interpreted

From the data collected during the course of this experiment, the researcher has found that punishment is more effective in altering behavior than positive reinforcement, but not by as clear a margin as expected. In the three test attempts, Subject A responded by eating both the lower-placed and the higher-placed food pellets as did Subject B. However, Subject A on average began eating the higher-placed pellets three minutes and 18 seconds earlier than he began eating the lower-placed pellets. Subject A also finished eating the higher-placed pellets nine minutes and 42 seconds earlier than he finished the lower-placed pellets. This contrasted sharply with Subject B, who began eating and finished eating both the higher- and lower-placed pellets at approximately the same time.

The hypothesis has been supported because the rat subjected

to punishment had memory of the bitter taste of the lower-placed pellets and avoided them temporarily in the test attempts. During the test attempts, the lower-placed pellets were not coated with the nail-biting deterrent, but the rat hesitated to eat them until it had exhausted the higher-placed bowl. In the three testing attempts, the rat had to sniff the lower-placed bowl before trying any of the pellets. The conditioning eventually gave way to the rat's investigation of the uncoated pellets, but it was enough to compel him to eat the higher-placed pellets first.

Subject A had eaten all the lower-placed pellets during the first and second conditioning attempts, but by the next few conditioning attempts, the rat had reduced consumption to zero of the coated pellets. This shows how punishment becomes more effective over time, and that it is more successful in altering behavior in environments where the negative contingency is ever-present. During the test attempts, punishment did have temporary effects on the behavior of Subject A, but these were paled in comparison to punishment's effects in the conditioning attempts. The presence of the aversive contingency made a large difference in punishment's effectiveness.

The data collected from Subject B's testing and conditioning showed a slight trend toward consumption of the higher-placed pellets before the lower-placed pellets, but the trend was not as definitive as Subject A's. At the beginning of the experiment, Subject B would eat all the lower-placed pellets before eating the higher-placed pellets, but this behavior changed a bit over time. During the later stages of the experiment, he began to eat from the higher-bowl first, but still ate from the lower-bowl before emptying the higher bowl. This slight shift was most likely caused by the fact that the rat was being reinforced to reach for the higher bowl. Subject B was receiving more food from it, so naturally this rat would develop a minor predilection for the higher bowl.

In conclusion, punishment caused Subject A to have a more highly-altered behavior in both the conditioning and attempting environments than did Subject B, which was subjected to positive reinforcement. Punishment had greater effects over time than it did at the commence of the experiment, and was more successful in the conditioning environment than the testing environment. Positive reinforcement did have some effects of the behavior of Subject B, but these effects were less obvious than those of punishment on Subject A.

Connection to Sources in Introduction

It makes sense that the researcher got these results because they were easily predictable from what he had read about punishment and positive reinforcement. Subject A avoided the lower-placed pellets because it had associated the bitter taste of the nail-biting deterrent with them (Heffner, n.d). During the testing, it hesitated to try these pellets, even when they were uncoated. Once it found that they were uncoated, however, it ate all of them. The fact that punishment does not permanently alter behavior was stated in Lawson and Watson's experiment, and was shown when Subject A eventually ate the lower-placed pellets (Lawson & Watson, 1963, p. 87).

As for positive reinforcement, Subject B was successfully conditioned to show some predilection for the higher bowl, but this was not as much as was to be expected. This rat should have associated the higher bowl with the larger supply of food, and therefore should have reached for it before going to the lower bowl. The positive reinforcer must have been weaker than the punishment, because it did not significantly increase the behavior of going to the higher bowl (Sherry, n.d.b).

Sources of Error Explained
While the rats where being observed, the researcher's presence may have slightly disturbed the rats and caused error in the data. The researcher had to sit next to the cages to record the rats' behavior, and the noise he made writing down the notes sometimes caused the rats to stop eating and look at him. In one particular instance, the researcher had particularly bothered Subject A by causing undue racket in attaching one of the bowls to the cage. The rat hid under his wheel and would not respond for several minutes.

The researcher also varied in how thickly he coated the

pellets with nail-biting deterrent. Before the first conditioning attempt he applied less solution to the pellets than he did in the other attempts, which may have caused Subject A to eat more of the lower-placed pellets during this first trial. This may have skewed the data for the first conditioning attempt, but most likely did not undermine the complete experiment.

In future experiments, the researcher should sit behind a piece of Plexiglas when observing the rats, so as not disturb them. Also, he should procure two more identical cages so that he can transfer the rats directly into a testing environment from the conditioning environment. This will prevent the rats from being bothered by the irksome process of installing the plastic bowls.

Connection to Real World Explained

At any given moment, the human is enveloped by multiple environments that range from the macrocosm of global and national society to the microcosm of the workplace and high school. These environments are constructed on the basis of promoting the growth, success, and productivity of the people they are composed of; as these environments have such a great influence, there is great debate as to whether punishment or positive reinforcement should be instituted in them (Dusak, n.d.). This experiment has proven that positive punishment is more effective than positive reinforcement, and that it should be instituted as a practical behavioral modifier. Even seemingly insignificant contingencies like bitter food have great effects on behavior and should be used by the creators of social environments.

Further Studies

If the researcher choses to continue this experiment next year, he will test reinforcement schedules rather than the effects of different behavior modifiers. Reinforcement schedules are classified as either continuous or partial, and vary in their effectiveness. Continuous reinforcement is defined as reinforcing a desired behavior each time it is practiced and partial reinforcement is defined as reinforcing a desired behavior only at certain times. The researcher's project will most likely deal with testing fixed-ratio and variable-ratio schedules. The two types differ in that fixed-ratio schedules reinforce desired behavior after a set amount of responses, and variable-ratio schedules reinforce desired behavior after a random number of responses (Sherry, n.d.d).

    Another possibility would be conducting the same experiment, but with a more intelligent animal, like a cat or a dog. The same setup could be used so long as it is mirrored on a scale large enough for the organisms being tested. It would be interesting to see if different species had the same responses as did the rats.

    The most innovative continuation, however, would be to apply to human subjects what has been discovered about positive reinforcement and punishment. This would be extremely difficult, as the approval process would take months and would likely result in the experiment being rejected. If in fact it was approved, the researcher would have to find volunteers who would agree to be submitted to punishment. Most likely, this punishment would be in the form of bitter and sweet-tasting food samples, distributed respectively for the completion of easy and difficult tasks.

Bibliography

Sources

Barker, B., Kreider, J., Peissig, J., Sokoloff, G., & Stansfield, M. (n.d.). *Stimuli.* Retrieved from http://www.psychology.uiowa.edu////.html

B.F. Skinner. (n.d.). Retrieved from NNDB website: http://www.nndb.com////

Doyle, O. (n.d.). *B.F. Skinner.* Retrieved from http://www.informationphilosopher.com////

Eckart, A., & Kentridge, R. W. (n.d.). *The Skinner Box.* Retrieved from Universitat Wurzburg website: http://genetics.biozentrum.uni-wuerzburg.de///.html

Educational Psychology Interactive. (n.d.). *An Introduction to Operant (Instrumental) Conditioning.* Retrieved from http://www.edpsycinteractive.org///.html

Farlex, Inc. (n.d.). *Operant Conditioning.* Retrieved from The Free Dictionary website: http://medical-dictionary.thefreedictionary.com/+conditioning

Heffner, C. L., Dr. (n.d.). Chapter 4: Learning Theory and Behavioral
    Psychology. In Pyschology 101. Retrieved from

AllPysch Online website: http://allpsych.com/psychology101/reinforcement.html

Lawson, R., & Watson, L. S., Jr. (1963, March). *Learning in the Rat {Rattus norvegicus) Under Positive vs. Negative Reinforcement with Incentive Conditions Controlled.* Abstract retrieved from The Ohio Journal of Science website: https://kb.osu.edu//////N02_087.pdf

Operant Conditioning Chamber. (n.d.). Retrieved from NationMaster.com website: http://www.statemaster.com/encyclopedia/conditioning-chamber

Rat Systematics. (n.d.). *Rat Species, Strains, Breeds and Varieties* . Retrieved from http://www.ratbehavior.org/.htm

Sherry, K. (n.d.b). *Introduction to Operant Conditioning.* Retrieved from http://psychology.about.com////cond.htm

Sherry, K. (n.d.c). *Pavlov's Dogs*. Retrieved from http://psycholog y.about.com////dogs.htm

Sherry, K. (n.d.d). *Schedules of Reinforcement* . Retrieved from http://psychology.about.com////.htm

Sherry, K. (n.d.a). *What is Behaviorism?* Retrieved from About.com website: http://psychology.about.com////.htm

Skinner, B. F. (1971). *Beyond Freedom and Dignity.* Indianapolis: Hackett Publishing Co.

Trubin, J. (2010, October). *From Pavlov to Skinner Box*. Retrieved from http://www.juliantrubin.com//.html

Dusak, J. (n.d.). Positive Reinforcement in the Workplace. Retrieved from
http://www.stfrancis.edu/content/ba/ghkickul/stuwebs/btopics/works/postrein.html

Nova Mycobacteriophage Now
Ji Woong Jung and Michael Mendelsohn

"We specifically got into microbiology and further phage biology from our teacher, Dr. Offner who has a keen interest in that field, even keeping close relations with leading phage scientists around the world. She started a club in which students can take up this interest in phages, more specifically mycobacteriophages for a science fair topic, and helps us understand the biology standpoint of them, and how to get started with our research. We applied the biology we learned throughout the year in figuring out the true question of our project and answering it with an unbreakable determination and insight. We learned to never give up."

Abstract:

Mycobacteriophages are viruses that infect and reproduce in host bacteria. Because of their abundance, recombination and mutations occur frequently, creating a diverse gene pool. Phages are grouped in clusters, which are groups of phages with similar structure, length, genes. We annotated and researched the phage Nova of Cluster D. Cluster D also includes the phages: Adjutor, Butterscotch, Gumball, PBI1, PLot, and Troll4. The intent of our project was to analyze the similarities between these phages, and their significance on an evolutionary scale. According to laws of genetics, the genome of organisms will be most similar to those most closely related; therefore, Nova's genome will be closest to other Cluster D phages. DNA Master's Blast feature was used in order to compare the genomes of phages.

The many similarities between the Cluster D phages occur in areas where the phages that have same genes, but are in different order. When a gene is present in only one phage, and if it has as Blast match to an unrelated bacteria, then it could be a product of recombination and gene insertion. Phages from Cluster H were the next most similar homologues to Nova after Cluster D. Nova was most similar to PLot in Cluster D, but in Cluster H, Konstantine was the most similar. Some of Nova's Blast matches suggested that Nova may have lived all over the world. Phages of Cluster D, interestingly enough were found in the Pennsylvania region, which could suggest that their common ancestor thrived in those regions. Through this we have also found medical research potentials.

Introduction:

This project analyzed the genomic sequence of the mycobacteriophage Nova and compared it to the sequencing of other mycobacteriophages as well as other organisms. Mycobacteriophages are viruses that infect bacteria. Phages are among the most numerous organisms on earth. They are interesting organisms to study because they reproduce so rapidly, thus making it possible to observe evolutionary changes in phages.

Question:

How does the genome of the mycobacteriophage Nova

compare to the genome of other mycobacteriophages and what evolutionary significance does it have?

Research:

Nova was a newly sequenced bacteriophage genome that had not yet been analyzed. The raw data for its genomic sequence was obtained from the University of Pittsburgh's phages database (phagesdb.org). Analyzing a newly sequenced, unannotated phage involves (1) genome annotation, which means identifying all the DNA and protein-coding genes, and (2) a comparative genomic analysis, which tries to understand the genomic architecture of the genome in the context of other phages and bacterial genomes. The researchers on this project learned about the structure of bacteriophages and the processes for analyzing their genes with the help of Dr. Susan Offner, an AP Biology teacher at Lexington High School.

In order to do the analysis, it is important to understand the structure and processes of mycobacteriophages. The general structure of a phage includes a capsid head enclosing the DNA, a portal connecting the capsid to the tail, a tail of alpha-helix proteins, and tail fibers.

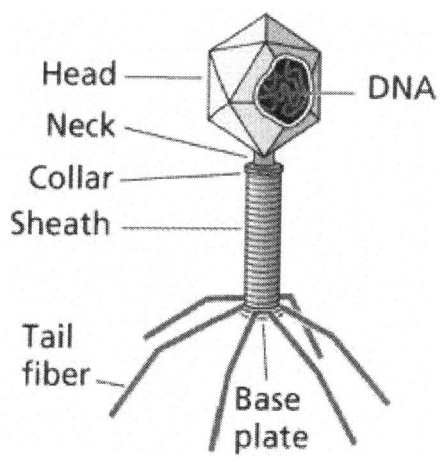

Head — DNA
Neck
Collar
Sheath
Tail fiber
Base plate

http://varnernic.glogster.com/chapter-12/

One of the important processes of phages is the process of how phages infect bacteria in order to reproduce. Phages infect bacterial hosts by binding their tail fibers to the specific receptor proteins of their hosts. Then they inject their DNA into the host cell, and then undergo either the lytic or lysogenic life cycles. During the lytic cycle, the phages use the host cell's mechanisms to produce 200-300 replicas of itself. The phage uses the bacteria's lysosomes to burst out of and kill the cell. During the lysogenic cycle, the phage similarly injects its DNA, but instead of reproducing immediately, it incorporates itself into the bacteria's circular DNA, forming a prophage. The phage's DNA is then replicated whenever the bacterial genome replicates. At right conditions, the phage DNA can remove itself from the prophage, and undergo the lytic cycle and reproduce. With 10^{31} phages in the world, and 10^{25} phage infections per second, there are countless insertions in which bacterial DNA fragments become incorporated into phage DNA. It is when this benefits the phage's fitness that it is selected for through many generations. In addition to insertion, mutations account for further genetic variation.

Mycobacteriophages are grouped into clusters, which are groups of phages with similar genomes, genes, gene number, structures, and base-pair lengths. The phages database classifies Nova as a part of cluster D. However, as part of this project, quantitative analysis was done to confirm its classification and to effectively compare it with its cluster. From the start, it was clear that Nova shared similar characteristics with other cluster D phages, including a base pair length of around 64-65 thousand base pairs, a similar shape, and a nearly identical raw genomic sequence.

Hypothesis:
The genes of Nova's genomic sequence are most similar to the genes of other phages in its cluster, and are most closely related to them evolutionarily.

Procedure:
The University of Pittsburgh's database of mycobacteriophages (phagesdb.org) was used as a resource to provide both information about annotated phages as well as their genomic sequences. The database provided the raw data for the newly sequenced phage called Nova. The aim of this project was to analyze the genomic sequence of Nova

and compare it to the sequencing of other organisms.

The analysis of Nova consisted of two phases. The first phase was genomic annotation which involved identifying where the genes which code for proteins in the DNA were. The first step was to obtain, from phagesdb.com; Nova's FASTA sequence which is the genomic sequence formatted by abbreviating nucleotides with A's, T's, G's, and C's. Once the FASTA was downloaded, it was saved as a word document.

A program called DNA Master was used to auto-annotate the sequences. Then, the genes were compared to all the other genes in phagesdb.org through a process called BLAST. Since the auto-annotation created an imperfect analysis, the researchers went through the data by hand to determine which sequences were correct.

Three main devices - the BLAST results, the ORF (open reading frame), and the Shine Dalgarno score – allowed the researchers to determine the start position of each gene. After the hand-analysis, the data was BLASTed once again. The aim of the genome annotation phase was to obtain the most definitive gene sequencing of Nova in order to compare it with gene sequences of other annotated phages as well as other organisms.

The second phase was the computational analysis in which three types of comparative tools were used to show similarities and differences between Nova and other phages; in particular, other cluster D phages. The three types of analytical tools were feature tables, gene maps, and dotter plots. The first chart created was the feature table which was a compilation of the second BLAST results. From the results, a table was produced with different columns showing the similarities, or homologue identities, to other phages, other out-of-cluster phages, non-phages, and any other interesting results found. The second step in the analysis was creating gene maps of Nova's genome. The gene maps for Nova were generated by DNA Master. Gene maps for the other cluster D phages were obtained from an outside source.

Three color-coded maps were created showing (1) the functions of Nova's genes, (2) the nearest homologue identity, and (3) the nearest non-cluster homologue. The nearest homologue identity is the actual gene found in the database with the most similar sequence to Nova's genes. The last form of analysis was the production of dotter plots. Dotter plots are graphs which show the similarities and differences between the unsequenced genome of two or more phages. These plots were generated by a program called J-Dotter. In order to create the dotter plots, it was necessary to download the FASTA of all phages which were to be included on the dotter plots. The FASTA files were copied and pasted into the program which ran the comparisons the phages genomes.

Analysis:

The central purpose of this study was the computational analysis which compared Nova's genomic sequence with other cluster D phages and non-cluster D phages.

The results displayed in the feature table show that 79 out of Nova's 87 genes had a nearest homologue identity to another cluster D phage. Not only were there a high number of matches, but the percent identity of the gene matches were mostly very high (95-100%). Of the cluster D phages, the cluster D phage named P Lot had the most homologue identities to Nova, sharing similar genes in 37 out of Nova's 87 genes. Other cluster D phages that had a significant number of homologues included Gumball (15 out of 87), Troll4 (13/87), PBI1 (12/87), and Adjutor (2/87). In addition to the high frequency and percent identity of Nova's gene in comparison to other cluster D genes, the gene map highlights certain key features of comparison between phages.

The second type of analysis was the gene maps. The gene maps show the position and function of some of the genes. The analysis of the gene maps showed that the position and length of most genes in Nova's genome were similar to the position and length of genes in the other cluster D phages. The gene map analysis also helped to learn about the function of some of Nova's genes. The gene map for the other cluster D phages labeled roughly 20 genes according to their function. Therefore, by comparing the gene maps of Nova against the other cluster D phages, it was possible to determine the function of some of Nova's genes.

The dotter plots were comparisons of the unsequenced genome between two or more phages. The control graph in this analysis was a dotter plot of Nova vs. Nova because its genome is identical to itself. The plot of Nova vs. the other cluster D phages was also run. From this data, it was discovered that Nova's genome was almost identical to each of the other cluster D phages with only a few discrepancies between them. There was one spot in which Nova was shown to have a different genomic sequence than all of the other cluster D phages.

Discussion:

We saw to this project two potential extensions for the future. One extension would be to create a phylogenetic tree of Cluster D phages, including Nova, showing those phages in evolutionary context of each other, seeing which share common ancestors and diverged more recently than others. We would use phylogeny DNA programs to produce this, and we can even go further into incorporating the cluster H phages as well.

Currently from just the blast matches of Nova's genes, we were able to see which Cluster D phages had the most similar number of genes to Nova, giving us a starting point for a potential phylogeny. PLot is the most similar to Nova with 37/87 nearest identity genes (DNA is most similar in those genes), then Gumball with 15/87, then Troll4 with 13/87, then PBI1 with 12/87, and finally Adjutor with 2/87. This could already suggest to us that PLot shares the closest common ancestor with Nova in the phylogeny, and the rest become less likely to have shared a common ancestor with Nova the less genes they match identity with. We can also look at the published genome maps of Cluster D phages, comparing their genes and gene order with each other, and to our map of Nova. A genomic map of a phage visualizes its entire genome, distinguishing genes and their base pair lengths from each other through rectangular blocks. We saw that although the cluster D phages all had a nearly identical number of genes (around 85-87), many existent in some were non-existent in others, and in different order in relation to that of the other cluster D phages. Yet with such subtle differences in their layouts, the phages still manage to pass their genes on to the next gene pool, and make their set of genes work for them. Yet the most conserved genes that code for crucial functions, were those that coded for functions Tape Measure, Minor Tail Subunit, Lys A and Lys B. Those genes were rock solid and prominent in similar location and length throughout the cluster D phages, because of their importance to all phages. We also noticed it quite unique that gene 27 of Nova had a closest non-cluster homologue match to Corndog, an un-clustered phage instead of a cluster H phage (cluster H was most prominent nearest non-cluster phage to Nova), which shows that in these phages' phylogeny and evolutionary paths, different clusters or groups of phages could have shared genes with Nova as it diverged from a common ancestor of cluster D phages, through infecting the same bacterial host. This adds to the further phylogeny potentials of an extension.

Another possible continuation to our project, and actually our main hopes and dreams for it, is to test for Nova's host range and its effects on phage therapy. What we have studied is Nova's individual genes, which mostly have interesting similarities and patterns of similarities to bacteria that are prominent and even deadly in the world today. We then looked into host range, which is the range or specific group of bacterial hosts that a certain phage is specific to in order to reproduce from. Like we said, phages can take up parts of their host's DNA when exiting the prophage state of their lysogenic cycle, which could effectively code for genes in the phage's new DNA that is beneficial and selected for. The blast matches we found and analyzed, suggests to us potential host range candidates to test for in the future, although some more than others. Gene 7 of Nova for example, does not have a nearest homologue identity match to any phage, but to bacteria called Syntrophobacter Fumaroxidans (with 46.3%). Because this gene has no similarity to any phages for that matter, but still seems to fill the spot for a gene that would regularly code for a digestive function called VIP2, it suggested to us that Nova could have diverged from a common ancestor of cluster D phages after a series of infections of bacteria like S. Fumaroxidans, that made it unique from others yet improved its fitness. In the dotter plot with Nova compared to Cluster D, there is a gap in the base pairs that code for gene 7 in Nova, further crediting this idea. Another interesting thing we found was a potentially prominent strain of bacteria that we could use to test for host range first in the future. This is called Mycobacterium marinum, and seemed highly likely to be in Nova and Cluster D phages' host range, because it had significant nearest non-mycobacteriophage homologue identity matches to genes of Nova around 20-30, that codes for the highest conserved proteins of all phages. The tape measure and minor tail subunit were among these proteins that M. Marinum was similar to, and because these proteins are so crucial in all phages, M. Marinum is suggested to us as a crucial part of Nova or cluster D phages' host range in the past that allowed it to take up DNA and diverge as a species. M. Marinum is actually bacteria that usually thrives in salt water conditions, and infects cuts as a skin disease, and it'll be interesting to see if Nova can kill these bacteria with certain genetic modifications in the lab. Other interesting finds were the blast matches of certain genes of Nova to Salmonella Enterica, Rhodococcus Opacus, Caldicellulosiruptor Hydrothermalis, Gordonia Bronchilais, Conexibacter Woesi, Chlamydia Muridarum, and Lactobacillus Vaginalis, all strains of bacteria that can be tested for Nova's host range in the lab. Some of these could tell us interesting things about

Nova's suitability in certain environmental conditions if these bacteria were in fact part of its host range. For example, Caldicellulosiruptor Hydrothermalis with a 65.6% identity to Nova's gene 20 is an anaerobic bacterium from Kamchatka thermal springs in Russia, which suggests to us that Nova or a common ancestor could have survived in hot fresh water or without oxygen. Another example is Conexibacter Woesei, an aerobic bacterium isolated from forest soil in Gerenzano, Italy, and is able to grow at 4 degrees Celcius and reduce nitrate to nitrite, which suggests to us that Nova can survive in oxygenated soil, and at low temperatures and high nitrogen concentrations. The others like Salmonella Enterica and Chlamydia Muridarum are deadly diseases around the world today, relating Nova to the world.

The way we would test to see which bacteria are in fact within Nova's host range, is to be genetically modified in many different ways (adding and subtracting certain genes), and using other cluster D phages as well, apply them onto cultures of different bacteria in testing. We would grow different bacteria on different agar plates, and introduce different phages and phage strains to different bacterial growths one at a time, until a circular clear plaque showing signs of bacterial death forms. If a certain phage thrives on these bacteria, it will infect outwards and could form concentric circles around the plaque of bacterial death. We look towards a special phage named Che12 as a model for finding cures in phage therapy. Che12 is a phage isolated from India, that is currently showing effect in infecting M. Tuberculosis, and further research can lead to it being a replacement or supplement to expensive antibiotics in eradicating M Tuberculosis in the future. Phage therapy is a great outlook on the future, because it is cheaper than antibiotics, easier to genetically engineer to resistant bacterial strains, and not harmful on humans, as the specific phages could target and kill of specific pathogens of the body.

Conclusion:
The hypothesis was originally that the genes of Mycobacteriophage Nova of cluster D, would be most similar to the genes of other phages in its cluster, and be most related to them evolutionarily. Through annotating our phage and producing a feature table from BLAST, our hypothesis was confirmed that Nova was most similar to the phages its cluster D in terms of genes. Our results analyzed with our knowledge on evolutionary biology strongly suggests to us that Nova is most evolutionarily related to the phages in its cluster. The importance of our project is that we become one step closer to understanding the ultimate tree of life, to the smallest critters that inhabit the world, and understand their use to impact the world.

Bibliography:
www.phagesdb.com
http://www.ncbi.nlm.nih.gov/
http://www.phage-biotech.com/links.html

Aragorn (**http://130.235.46.10/ARAGORN/**)

tRNA Scanner (**http://lowelab.ucsc.edu/tRNAscan-SE/**)

http://expasy.org/

http://www.plosgenetics.org/

http://www.sciencedirect.com

Microsoft Word and Excel

Google Images

DNA Master Program

Gene Mark Program

J-Dotter Program

Ink-Scape Program

FedEx Kinko's Printing

Phage Hunting Program (11.12.2010) Instructional Guide

Comparative Genomic Analysis of Mycobacteriophage Tweety…etc.

Viruses

Mycobacteriophage Clusters (Feb.2009)

Filtered Results

Acknowledgements:

We would like to thank these following people for their undying support and love. Dr. Offner thought us the background of our topic, and kept us intrigued and interested the whole way, and we thank her greatly for that. An unforgettable appreciation to our parents, Young Jung, Kyung Kim (Philip), and Noah Mendelsohn, Eileen Jay (Michael), who not only drove us, and fed us, and watched over us, but reminded us that there's no stronger force for success than family.

Impact Concussion
By Elizabeth Kawa and Bishop Feehan

"First of all, I want to go to medical school so I wanted to do a project based on medicine. My sister is on a soccer team where half of the players wear headgear. Knowing that concussions are becoming a "hot topic" in the world of sports, I wanted to see if headgear was effective in reducing the impact of concussion. The first thing I did was order Styrofoam manikin heads and I built my project around that. My mom, who is a nurse practitioner, guided me in the research. Also, one of the experts in Home Depot provided me with tips and materials for building my head mount. I approached her with my project ideas and she very willingly assisted me with structure. Through a lot of trial and error, I have found that headgear is indeed successful in decreasing the impact of concussion."

Abstract

Purpose: The purpose of this experiment is to discover whether the degree of traumatic brain injury, when suffered by an athlete, is lessened by protective headgear.

Problem: Does protective headgear lessen the degree of traumatic brain injury as measured through the depth of the spike produced on the graph paper?

Hypothesis: If an athlete wears protective headgear during a sport and suffers traumatic brain injury, then the impact of traumatic brain injury will be lessened as measured through the depth of the spike produced on the graph paper.

Concise Description of Experiment: A Styrofoam head was placed on a coil in front of a wooden backboard, to which a piece of graph paper was attached. A marker was inserted into the back of the Styrofoam head. A ball was placed into a catapult and propelled at the bull's-eye on the forehead of the Styrofoam head multiple times at the same speed and distance so that the marker produced spikes on the graph paper. The depth of the spikes determined the degree of traumatic brain injury. This was tested without protection, with Full 90 headgear, and with a layer of rubber.

Expected Outcome: If an athlete wears protective headgear during a sport and suffers traumatic brain injury, then the impact of traumatic brain injury will be lessened as measured through the depth of the spike produced on the graph paper.

Significant Results: This researcher found that the average depth in the spikes was larger when there was no protective padding. Full 90 headgear may serve as a protective padding that does lessen the impact of traumatic brain injury as evidenced through this experiment. The layer of rubber may prove to be an effective source of protection as well.

Introduction
Rationale:

The topic of concussions is very timely in society as the rate of teens who suffer from this traumatic brain injury continues to increase. More than ever, research is being done on the ramifications of repeated concussions in athletes. Full headgear is used in sports, most specifically in soccer. Headgear may provide a false sense of security to athletes and their parents. This researcher wanted to determine if this protection is indeed successful in lessening the impact of traumatic brain injury, thereby, decreasing the long term effects of concussions.

Background:

Sports-related concussions, a common injury that is often neglected by pediatric and adolescent athletes, is becoming a "hot topic" among the media and medicine (Halstead & Walter, 2010). In simple terms, a concussion is a temporary loss of normal brain function (Durani, 2009). It may be caused either by a direct blow to the head, face, or neck or elsewhere on the body with an intense force transmitted to the head. Some may also refer to this as a "mild traumatic brain injury," or mTBI for short (Halstead & Walter, 2010). One of the most common reasons that people suffer from concussions is through a sports injury, especially in high-contact sports such as football, boxing, hockey, basketball, and soccer (Durani, 2009).

There are four categories under which the signs and symptoms of a concussion fall: physical, cognitive, emotional, and sleep. The most frequently reported symptom is a headache (Halstead & Walter, 2010). Because of the lack of understanding of a concussion, athletes may not recognize that they are experiencing concussive symptoms. This may put them at a risk for serious injury if they return to a game before they should (Information about Concussion for Kids and Teens, 2010). Symptoms may not appear until several hours after the head injury occurs. Some of these include memory loss, nausea, headaches, blurred vision, slurred speech, difficulty concentrating, thinking or making decisions, and difficulty with coordination or balance. There are two different types, or grades, of a concussion: the simple concussion and the complex concussion. With a simple concussion, the symptoms will return to normal over a period of seven to ten days. With a complex concussion, symptoms can last for over ten days (Durani, 2009).

Not only is the popularity of soccer unquestioned, but also the head injuries that occur in soccer as well (Concussions Among University Football and Soccer Players, 2010). Currently, in the United States, there are well over three million children and adolescents that play on high school or youth soccer teams. As of now, the participation rate is increasing by about 21% annually. As stated before, soccer is among the sports with the highest risk of concussion. According to Ofer Zur of the Zur Institute, concussions make up 20% of soccer injuries (Zur, 2010). Just as the act of heading has not yet been proven to cause concussions, the effects of years of heading are still under debate. Other types of impact include head to ground, head to goal post, and head to head, which are also known to cause concussions and long term negative consequences (Concussions Among University Football and Soccer Players, 2010).

A Norwegian study of adult soccer players, who had begun playing in their youth, presented mild to severe deficits in attention concentration and memory. This was true for 81% of the players that were tested. Those players who headed the ball more frequently had higher rates of cognitive loss. A Dutch study also presented similar results. During another study, which included a team of university teachers from both the School of Medicine and the School of Engineering and Applied Science, concluded that the most critical danger in heading is the linear and angular accelerations with changes in velocity over time. The brain experiences a shear force that causes deformation in areas of the brain. These areas house the penetrating blood vessels and the hemispheres that come in contact with the brain. Depending on the velocity and the random amount of acceleration, they believe this can occur with mild impact (Zur 2010)

As indicated by the U.S. Consumer Product Safety Committee, which collects data on injuries in all sports, there has been an estimate of over one thousand soccer-related concussions in the United States every year. Consistent with the data provided by the Sporting Goods Manufacturers Association and the CPSC, the rate of concussions has grown nearly twice as fast as the rate of growth of soccer players over the past eight years. Reported concussions in soccer are caused by head to player contact (40%), head to the ground or goal post (10.3%), head to soccer ball (12.6%), and not specified (37%) (Concussions Among University Football and Soccer Players, 2010).

There are two types of forces present during the impact that can lead to concussions, linear force and rotation force. Linear force is experienced when the body is pushed without any twisting or tilting. For example, a wind that blows directly onto the back exerts a force that is almost entirely linear. Rotational acceleration is experienced when the body is struck by a blow that tends to turn or tilt the body. For example, someone bumping into one of the shoulders from behind generates a rotational force on the body by turning it around (Concussions Among University Football and Soccer Players, 2010).

One of the issues associated with concussions is that of repeated concussions within a fairly short time frame (Information about Concussion for Kids and Teens, 2010). It is currently believed that a person who has suffered from one concussion is four to six times more likely to be diagnosed with another concussion. According to tests performed by Full 90 Sports, "concussive-level impacts are reduced to sub-concussive levels when the headguard is worn, with the reductions of upwards of 50% to both linear and rotational acceleration in impacts typical in soccer." The results of these tests show that headgear in soccer can be worn without altering the game (Concussions Among University Football and Soccer Players, 2010). Full 90 headgear is a device made of shock-absorbing foam situated between an outer layer of Lycra and an inner layer of sweat-absorbing polypropylene (Full 90: Stay in the Game). Many athletes, though, object to wearing the headgear because they claim that it is a "dorky" fashion statement. Parents may also be unwilling to pay the cost of the headgear which can range from twenty-five dollars to forty dollars (Reider).

Experiments conducted at North Dakota State University by Dr. Mariusz Ziejewski investigated whether head to ball impact was adequate to cause injury, deliberate heading versus the state of being unprepared for the impact. He performed computer stimulations based on impact data, determining that "ball impacts can cause concussion if the head is not prepared for impact, impact forces can be reduced by protective padding, Full 90 headgear reduced the strain on the brain by 50%, and the linear acceleration in a heading incident was reduced by 40% by Full 90 in typical collisions" (Concussions Among University Football and Soccer Players, 2010).

Rubber may serve as a protective padding for the impact that may cause mild traumatic brain injury. Rubber exists in two forms, natural and synthetic, that are very common in today's modern society. Its flexibility makes it suitable for various kinds of shock absorbers, designed to absorb the pressure and impact. Rubber shock absorbers are used widely in automobiles acting as a protection against vibration and shock. They absorb large amounts of kinetic energy which are generated from impact. The machine rubber shock absorber also absorbs the kinetic energy and prevents the machine from excessive shocks and recoil forces. In this way, rubber may be a good choice of material to be fabricated in headgear (Shock Absorbers World).

Through this experiment, this researcher is attempting to discover whether protective headgear lessens the degree of traumatic brain injury when suffered by an athlete. The hypothesis is "If an athlete wears protective headgear during a sport and suffers traumatic brain injury, then the impact of traumatic brain injury will be lessened as measured through the depth of the spike produced on the graph paper." A Styrofoam head will be placed on two coils in front of a wooden backboard, to which a piece of graph paper will be attached. A marker will be inserted into the back of the Styrofoam head. A ball will be placed in a catapult and propelled at the bull's-eye on the forehead of the Styrofoam head multiple times at the same speed and distance. The marker will produce spikes on the graph paper when the coils flex back from the impact. This will be tested without protection, with Full 90 headgear, and with a thin layer of rubber.

Materials and Method

→ Materials:
- 2 pieces of 2x3x7 ½ inch wood
- 3 pieces of 2x3x14 inch wood

- 2 pieces of 2x3x24 inch long wood
- 1 piece of 2x3x28 inch long wood
- 1 piece of 2x3x10 ½ inch long wood
- plastic cup
- 2 large hooks
- long screws
- nails
- 8 oz can orange paint
- 8 oz can red paint
- paintbrush
- 24 inch bungee cord
- 1x12 inch dowel
- 2 ½x6 inch piece of thin rubber
- Full 90 Headgear
- Styrofoam head
- red marker
- green marker
- blue marker
- tape measure
- saw
- screw driver
- 2 ¼ inch washers
- 2 coils
- wiffle ball
- 12 pieces graph paper
- ruler
- masking tape
- 2 boards of 24x24x1 inch wood

→ Controls:
- distance between head mount and catapult
- catapult
- head mount
- waffle ball
- markers

→ Variables:
- distance arm of catapult was pulled back

- impact
- position **of** impact on head
- position **of** head

→ Experimental Setup and Directions:

First, this researcher built a catapult. All wood was 2x3 inches. Two pieces were cut to 7 ½ inches, three pieces were cut to 14 inches long, another two pieces were cut to 24 inches, one piece was cut to 28 inches long, and one piece was cut to 10 ½ inches long. This researcher took one of the 24 inch pieces and measured 3 ½ inches from the end. A mark was made to represent the pivot hole. A hole was drilled here so that the wooden dowel would slide through. The same process was done on both 24 inch pieces. Then, this researcher made marks at 5 ½ inches and 8 inches. Lines were drawn to connect these marks. A saw was used to cut down on both the lines so the cut went halfway through the 2x3 piece of wood. A chisel and hammer were used to cut out the excess wood.

This process was repeated for the other 24 inch piece of wood. This process was also done to two of the fourteen inch pieces. This researcher only did one end of each piece. A line was made at 2 and 2 ½ inches from the end and the saw cut halfway through the wood. This researcher chiseled away the excess wood. Because a notch was cut into the 24 inch piece and a notch was cut into the 14 inch piece, the two pieces of wood fit together nicely. This researcher then screwed the 24 inch pieces and the 7 ½ inch pieces together.

The notches that were cut in the wood were put so that they were facing the outside. One of the eye hooks was screwed into the end.

This researcher screwed the two upright pieces into place. Then, the 10 ½ inch piece was screwed across the top of the two upright pieces of wood. The final 14 inch piece of wood was custom cut so that the ends were 45° angles. This researcher screwed it directly into place. This piece is important because it adds a lot of strength to the catapult. This researcher took the 28 inch piece and drilled a hole into the end so that the wooden dowel would slide easily through. A small plastic cup was attached to the other end of the wood with two nails. The other eye hook was screwed into place. This researcher slid one end of the wooden dowel into one end of the catapult, through the swinging 28 inch piece of wood, and then through the other end of the catapult. Lastly, the bungee cord was attached to the eye hooks. The assembled catapult was painted orange and red. This completed the catapult.

This researcher then nailed two 24x24x1 inch boards of wood at a 90° angle. The head mount was painted orange and red. On the backboard, a sheet of 8x11inch graph paper was attached using masking tape. Two coils were bolted to the bottom board using two ¼ inch washers, two nails, and a hammer. The coils were nailed 4 ¼ apart and 12 inches from the backboard. Two holes were carved in the bottom of the Styrofoam head. This researcher fitted the Styrofoam head onto the coils. This completed the head mount.

The catapult was placed 36 ¾ inches away from the Styrofoam head located on the wooden board. A plastic ball was placed in the plastic cup. A bull's-eye was marked on the Styrofoam head and the layer of rubber. The headgear already had a Full 90 bull's-eye. The 28 inch piece of wood, acting as an arm, was pulled back for a total of 7 inches. The wiffle ball was propelled at the bull's-eye on the unprotected Styrofoam head 10 times. This researcher then fitted the Full 90 headgear onto the Styrofoam head. The ball was again propelled at the head 10 times. This researcher removed the headgear and attached a piece of thin rubber to the head using masking tape. The ball was propelled at the head 10 times. This researcher made sure the head was erect after each impact with ball. This experiment in its entirety was repeated for three more occasions.

Results

→ Tables and Graphs:
→ Written Descriptions:

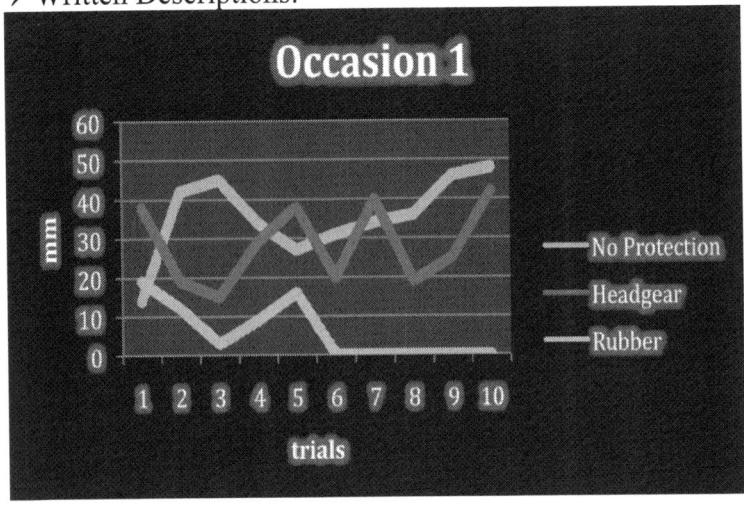

Upon reviewing the spikes from occasion one, it appears that, overall, those made when the head was unprotected were larger than those made when the head was wearing headgear and rubber. There is a significant decrease in the depth of the spikes when the rubber was attached to the Styrofoam head. As reflected in the graph, when the head was fitted with the thin layer of rubber, the coils did not flex back enough to make spikes on the graph paper for a total of five trials.

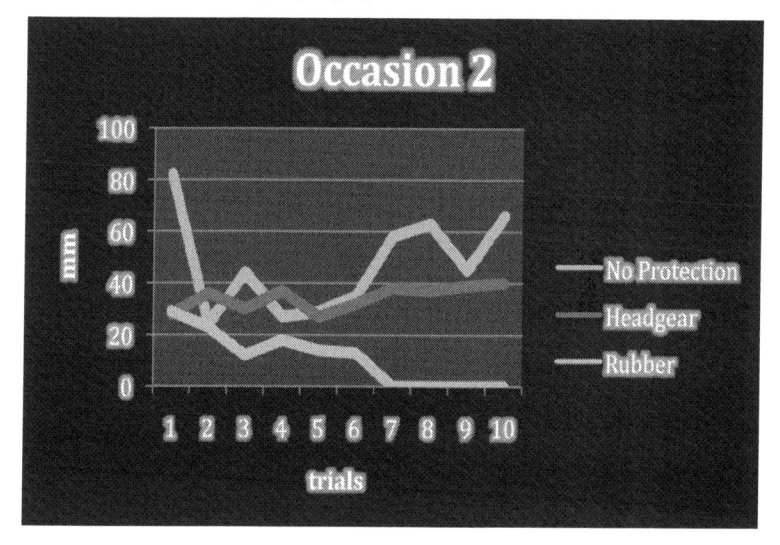

Occasion 2

The depth in the spikes ranged from zero millimeters to eighty-two millimeters. The largest depth was at 82 mm and was made when the Styrofoam head was unprotected. The depths of the spikes when the head is unprotected are larger than those made when the head is protected with headgear or rubber. As compared to occasion one, there is still a significant decrease in the depth of the spikes when there is rubber attached to the head. During occasion two, there were four trials where the coils did not flex back enough to make a mark on the graph paper.

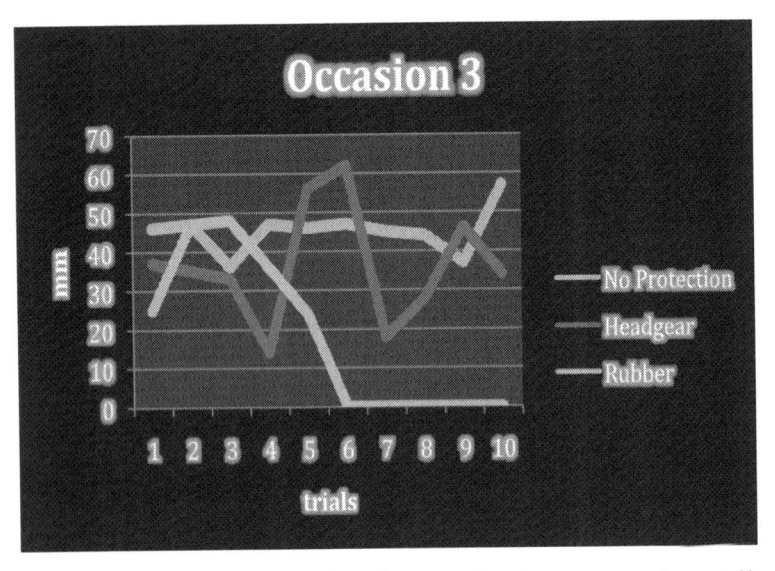

Occasion 3

[Graph showing mm (0-70) vs trials (1-10) with three lines: No Protection, Headgear, Rubber]

The largest spike was made when the head was wearing Full 90 headgear with a depth of 62 mm. However, it provided spikes with two of the lowest depths. During occasion three, the results presented when the head was wearing Full 90 headgear are less consistent as compared to occasions one and two. As depicted in the graph, the depths of the spikes when the head was fitted with rubber are greater as compared to occasions one and two. There were five trials, though, where the coils did not flex back enough to create a spike on the graph paper.

Occasion 4

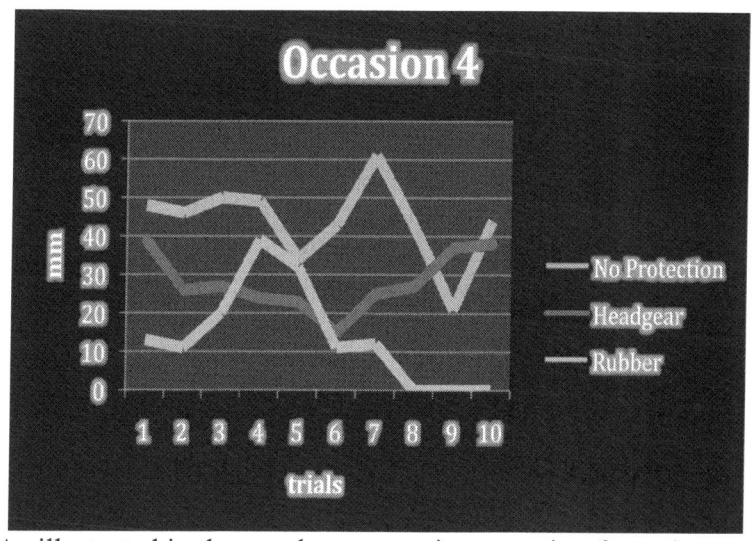

As illustrated in the graph representing occasion four, the depths of the spikes when the head was unprotected are again greater than those made when the head was protected with either headgear or rubber. The largest spike, made when the head was unprotected, had a depth of 61 mm. As one can see, the depths of the spikes made when the head was protected by Full 90 headgear are much smaller in general, ranging from fifteen to thirty-nine mm. When the rubber was attached to the head, there were two trials where the depths of the spikes were greater than those of the headgear and again there were only three trials where there were no spikes made on the graph paper.

Conclusion

→ Connection to Sources:
→ Results Interpreted:
→ Sources of Error:
→ Connection to Real World:
→ Further Studies:

Concussions are a "hot topic" among athletes in society. In simple terms, a sports-related concussion is a temporary loss of brain function, often neglected by pediatrics and young athletes. A concussion, also known as a mild traumatic brain injury, is most commonly caused by high-contact sports such as basketball, hockey, and football. Soccer is also among the sports with the highest risk of concussion. Concussions make up about 20% of soccer injuries. According to tests performed by Full 90 Sports, a concussive headgear in soccer can be worn without altering the game and results showed that impact forces can be reduced by this protective padding.

If an athlete wears protective headgear during a sport and suffers traumatic brain injury, then the impact of traumatic brain injury will be lessened as measured through the depth of the spikes produced on the graph paper. This researcher found that when the depths of the spikes from the four occasions were averaged, that the depths made from the unprotected Styrofoam head were greater than both those made from the head that was wearing headgear and those made when the head was fitted with rubber. One may take this a step further and conclude that the impact was greater without protection and therefore, may result in a more severe injury, or a higher grade of traumatic brain injury. In addition, it is significant to note that when the Styrofoam head was protected with the layer of rubber, the coils did not flex enough for the marker to create a spike on the graph paper during several trials. This may reflect the shock absorbing aspect of rubber. This may lead to the question of whether rubber should be incorporated into the manufacture of headgear. During occasion one, the depths of the spikes without protection as compared to those with Full 90 headgear decreased by a total of 20%. From no protection to the layer of rubber, the depths of the spikes decreased by a total of 67%. During occasion two, the depths decreased by a total of 27% from without protection to headgear and by a total of 61% from no protection to the rubber. During occasion three, the depths of the spikes decreased respectively by 20% and then by 20% again. On occasion four, the depths decreased by a total of 36% and then by a total of 55%. These results are consistent with studies done by Full 90 sports, where the headgear was reported to have reduced strain on the brain by 50%.

In the process of this scientific experiment, this researcher attempted to have the impact be as consistent as possible. Because of this, the results in the graph were consistent. This was done through precise measurements between the position of the catapult and the position of the head mount. The arm of the catapult was pulled back to the same distance with every trial. Only those impacts were counted that hit the bull's-eye on the Styrofoam head. The markers that were used were exactly the same brand and size. They were placed in the same crevice in the back of the Styrofoam head. However, there still may have been human error, which could have affected the results. The impact may have been slightly varied. For example, the impact of the waffle ball may not have hit the bull's-eye in its direct center on the Styrofoam head during certain trials. This is evidenced on occasion three when the largest spike was made when the Styrofoam head was protected by Full 90 headgear.

Despite the limitations, this experiment can have important consequences in the world of sports. If this researcher was to repeat this experiment, she would incorporate pieces of rubber at different sites within the Full 90 headgear and the impact would be retested. Although this researcher attempted to make the results as consistent as possible, the experiment could be improved by using an automatic ball pitcher. This may make the direction, speed, and impact even more reliable. The experiment could be trialed on different areas of the head, such as the occipital, temporal, and parietal regions. Different degrees of impact could also be trialed. The results may be more pertinent if a sturdier manikin head is utilized as opposed to a Styrofoam head. These modifications in the experiment may make the results more relevant to mild traumatic brain injury today.

Should headgear be made mandatory in high school

and college soccer? Maybe one of the answers to this question is to make headgear a required part of the soccer uniform, just as the helmet is a required part of the uniform in football. This researcher concluded that the presence of Full 90 headgear does decrease the impact of concussion. Because the rubber proved to be an efficient source of protection, the headgear may be made more effective by including a thin layer of rubber into the construction of the Full 90 headgear. High school and college administrations should take part in enforcing the rules about headgear and parents and teens should be more educated regarding the long term ramifications of concussions. Teen athletes should save their smarts by wearing headgear.

Bibliography

Concussions Among University Football and Soccer Players. 2010. 17 October 2010
<www.full90.com>.
Durani, Yamini. Concussions. April 2009. 12 September 2010
<http://kidshealth.org>.
Full 90: Stay in the Game. 2010. 24 January 2010

<http://www.full90.com/products/protect/premier/>.
Halstead, Mark E and Kevin D Walter. "Clinical Report- Sport-related Concussion in   Children andAdolescents." Pediatrics: Official Journal of the American Academyof   Pediatrics (2010): 597-615.
Information about Concussion for Kids and Teens. 2010. 17 October 2010

<http://thechildrenshospital.org/conditions/rehab/concussion/kids_info.aspx>.

Reider, Jacob. <u>Soccer Headgear.</u> 29 March 2009. 24 January 2010 <http://www.docnotes.com/2009/03/soccer-headgear.html>.

<u>Shock Absorbers World.</u> 2010. 24 January 2010
    <http://www.shockabsorbersworld.com/rubber-shock-absorber.html>.

Zur, Ofer. <u>Heading in Soccer.</u> 11 September 2010. 12 September 2010
    <http://www.zurinstitute.com/headinginsoccer.html>.

List of Science Fair Contacts by State

The local and statewide fairs serve as the main feeder system for the National Intel Science and Engineering Fair system. To get involved, reach out directly to the contacts shown below, and they can give you specific information about how their particular system works. In some, you need to compete first at the local and regional level, in order to participate. Others allow people to enter directly, or will make accomdations in the event that your secondary school doesn't offer a science fair program.

Again these only refer to the ITEF fair, but there are many others such as the Siemens Fair that you can learn more about just by googling them directly.

ALASKA
Anchorage
Alaska Science and Engineering Fair
http://www.alaskasciencefair.org
Juneau
Southeast Alaska Regional Science Fair
165 Behrends Ave., Juneau, 99801

ALABAMA
Auburn
Greater East Alabama Regional Science and Engineering Fair

http://www.auburn.edu/academic/science_math/cosam/outreach/

Birmingham

Central Alabama Regional Science and Engineering Fair

http://www.uab.edu/cord

Huntsville

North Alabama Regional Science and Engineering Fair

http://narsef.uah.edu

Huntsville

Alabama State Science and Engineering Fair

http://www.uah.edu/ASEF

Livingston

West Alabama Regional Science Fair

http://sciencefair.uwa.edu

Mobile

Mobile Regional Science Fair

http://www.southalabama.edu/mrsef/

ARKANSAS

Arkadelphia

South Central Arkansas Regional IV Science Fair

http://www.dawson.dsc.k12.ar.us

Conway

Arkansas State Science Fair

http://www.uca.edu/org/assfa/

Fayetteville

Northwest Arkansas Regional Science and Engineering Fair

http://cmase.uark.edu

Hot Springs
West Central Regional Science fair
http://sites.google.com/a/asmsa.org/wcrsf/
Jonesboro
Northeast Arkansas Regional Science Fair
http://altweb.astate.edu/neasf/
Little Rock
Central Arkansas Regional Science Fair
http://www.lrcentralhigh.org/carsf.htm
Magnolia
Southwest Arkansas Regional Science Fair
http://sites.google.com/site/ctemssau
Monticello
Southeast Arkansas Regional Science Fair
http://www.uamont.edu/FacultyWeb/Nelson/RSF/

ARIZONA
Phoenix
Arizona Science and Engineering Fair
http://azsef.asu.edu
Sierra Vista
SSVEC's Youth Engineering and Science Fair
http://www.yesfair.com
Tucson
Southern Arizona Regional Science and Engineering Fair
http://www.sarsef.org

CALIFORNIA

Contra Costa
Contra Costa County Regional Science Fair
http://www.cccsef.org
Costa Mesa
Orange County Science and Engineering Fair
http://www.ocsef.org/
Fresno
Central California Regional Science, Mathematics and Engineering Fair
http://www.fcoe.org/smefair
Livermore
Alameda County Science and Engineering Fair
http://acsef.org
Los Angeles
Los Angeles County Science and Engineering Fair
www.lacoe.edu/sciencefair
Los Angeles
California State Science Fair
http://www.usc.edu/CSSF
Palos Verdes Peninsula
Palos Verdes Peninsula Unified School District Science and Engineering Fair
http://www.pvphs.com
Sacramento
Sacramento Regional Science and Engineering Fair
http://www.srsefair.org
San Bernardino
Riverside, Inyo, Mono, San Bernardino Science and

Engineering Fair
San Diego
Greater San Diego Science and Engineering Fair - 1
http://www.GSDSEF.org
San Francisco
San Francisco Bay Area Science Fair, Inc.
http://www.sfbasf.org
San Jose
Santa Clara Valley Science and Engineering Fair Association
http://www.science-fair.org
Santa Cruz Santa Cruz Science Fair
http://science.santacruz.k12.ca.us
Seaside
Monterey County Science and Engineering Fair
http://www.montereycountysciencefair.com

COLORADO
Alamosa
San Luis Valley Regional Science Fair, Inc.
Boulder
Roche Colorado Regional Science Fair
http://www.bvsd.org
Brush
Morgan-Washington Bi-County Science Fair
Colorado Springs
Pikes Peak Regional Science Fair
http://www.pprsf.org
Denver

Denver Metropolitan Regional Science and Engineering Fair
http://ahec.ucdenver.edu/sciencefair/
Durango
San Juan Basin Regional Science Fair
http://www.sjboces.org/sciencefair
Fort Collins
Colorado Science and Engineering Fair
http://www.csef.colostate.edu
Grand Junction
Alpine Bank Western Colorado Science Fair
Western Colorado Regional Science Fair
Greeley
Longs Peak Science and Engineering Fair
http://mast.unco.edu/science_fair/
La Junta
Arkansas Valley Regional Science Fair
http://www.ojc.edu/ScienceFair.aspx
Sterling
Northeast Colorado Regional Science Fair
Connecticut
Danbury
Science Horizons, Inc. Science Fair and Symposium
http://www.sciencehorizons.info
Hamden
Connecticut State Science Fair - Life Sciences
http://www.ctsciencefair.org
District of Columbia
Washington, DC

District of Columbia Science and Engineering Fair
http://www.dcsciencefair.com

FLORIDA

Brooksville
Hernando County Regional Science and Engineering Fair
http://hernandoschools.org/index.php?option=com_rubberdoc&view=category&id=50&Itemid=260

Bushnell
Sumter County Regional Science Fair
http://www.sumter.k12.fl.us

Fort Myers
Thomas Alva Edison Kiwanis Science and Engineering Fair
http://www.edisonfairs.net

Fort Walton Beach
Northeast Panhandle Regional Science and Engineering Fair
http://www.okaloosa.k12.fl.us/science/

Ft Lauderdale
Broward County Science Fair

Gainesville
Alachua Region Science and Engineering Fair
http://tlc.sbac.edu/scifair/

Green Cove Springs
Clay Rotary Regional Science and Engineering Fair
http://www.clay.k12.fl.us

Inverness
Citrus Regional Science and Engineering Fair

Jacksonville

Northeast Florida Regional Science and Engineering Fair

http://www.nefrsef.org

Jensen Beach

Martin County Regional Science and Engineering Fair

Kissimmee

Osceola County Schools Science and Engineering Fair

Lake City

Suwannee Valley Regional Science and Engineering Fair

Lakeland

Polk Region Science and Engineering Fair

Melbourne

South Brevard Science and Engineering Fair

Merritt Island

Brevard Intracoastal Regional Science and Engineering Fair

Merritt Island

Brevard Mainland Regional Science and Engineering Fair

Miami

South Florida Science and Engineering Fair

http://science.dadeschools.net/scienceFair/default.html

New Port Richey

Pasco Regional Science and Engineering Fair

Ocala

Big Springs Regional Science Fair

www.marion.k12.fl.us

Orlando

Dr. Nelson Ying-Orange County Science Exposition - Physical Science

http://www.yingexpo.com

Orlando
State Science and Engineering Fair of Florida - Ying Scholars
http:\\www.floridassef.net
Palmetto
Manatee Lockheed Regional Science and Engineering Fair
http://www.manatee.k12.fl.us/curriculum/science/sciencehome_files/Page643.htm
Pensacola
West Panhandle Regional Science and Engineering Fair
Saint Augustine
River Region East Science Fair
http:\\www.stjohns.k12.fl.us
Sanford
Seminole County Regional Science, Mathematics & Engineering Fair
Sarasota
Sarasota Regional Science, Engineering and Technology Fair
http://www.sarasota.k12.fl.us
Sebring
Heartland Regional Science and Engineering Fair
Seminole
Pinellas Regional Science and Engineering Fair
Tallahassee
Capital Regional Science and Engineering Fair
http://www.crsef.org
Tampa
Hillsborough Regional Science Fair

http://www.sdhc.k12.fl.us

Tavares

Lake Regional Science & Engineering Fair

Vero Beach

Indian River Regional Science and Engineering Fair

http://www.edfoundationirc.org

West Palm Beach

Palm Beach County Science and Engineering Fair

http://www.palmbeachschools.org/sc/SciFair2010/Science_and_Engineering_Fair.asp

GEORGIA

Albany

Darton College/Merck Regional Science Fair

Athens

Northeast Georgia Regional Science and Engineering Fair

Athens

Georgia State Science and Engineering Fair

http://www.georgiacenter.uga.edu/oasp

Atlanta

Atlanta Science and Mathematics Congress

http://www.atlantapublicschools.us/mathandscience/

Brunswick

Coastal Georgia Regional Science and Engineering Fair

http://www.ccga.edu/ScienceFair/

Buford

Gwinnett Regional Fair

http://gwinnettsciencefair.com/

Conyers
Rockdale Regional Science & Engineering Fair
Decatur
Dekalb Science & Engineering Fair
http://www.dekalb.k12.ga.us/
Griffin
Griffin RESA Regional Science Fair
http://www.griffinresa.net/2009ScienceFair.pdf
McDonough
Henry County Science and Engineering Fair
http://www.henry.k12.ga.us
Milledgeville
Georgia College & State Univ. Regional Science and Engineering Fair
http://chemphys.gcsu.edu/~science
Savannah
Savannah Ogeechee Regional Science and Engineering Fair
http://www.gtsavrsef.org
Warner Robins
Houston Regional Science and Engineering Fair

HAWAII
Hilo
Hawaii District Science and Engineering Fair
Honolulu
Hawaii Association of Independent Schools Science and Engineering Fair
Honolulu

Hawaii State Science and Engineering Fair

http://www.hawaii.edu/acadsci

Kailua

Windward District Science and Engineering Fair

Lihue

Kauai Regional Science & Engineering Fair

Pearl City

Leeward District High School Science Fair

WailukuMaui Schools Science and Engineering Fair

IOWA

Ames

State Science and Technology Fair of Iowa

http://www.sciencefairofiowa.org

Cedar Rapids

Eastern Iowa Science and Engineering Fair

http://www.eisef.org

Sheldon

Western Iowa Science and Engineering Fair

ILLINOIS

Arlington Heights

Illinois Junior Academy of Science North Suburban Region 6 Science and Engineering Fair

http://region6ijas.weebly.com/

Chicago

Chicago Public Schools Student Science Fair

http://www.chicagostudentsciencefair.org

DeKalb

Illinois Junior Academy of Science Region V Science and Engineering Fair

http://www.niu.edu/clasep/fairs/science/index.shtml

Edwardsville

Illinois Junior Academy of Science Region XII Science Fair

Macomb

Heart of Illinois Science and Engineering Fair

Springfield

Illinois Junior Academy of Science Region X Science Fair

INDIANA

Angola

Northeastern Indiana Tri-State Regional Science Fair

http://www.neitssciencefair.org/

Bloomington

South Central Indiana Regional Science and Engineering Fair

http://www.indiana.edu/~oso/SciFair/index.htm

Evansville

Pott Foundation Tri-State Regional Science and Engineering Fair

http://www.usi.edu/science/fair

Fort Wayne

Northeast Indiana Regional Science and Engineering Fair

http://www.ipfw.edu/scifair/

Gary

Calumet Regional Science Fair
http://www.calumetregionalsciencefair.org/

Greencastle
West Central Indiana Regional Science and Engineering Fair
http://www.sefi.org/wcir

Hanover
Southeastern Indiana Regional Science Fair

Indianapolis
Central Indiana Regional Science and Engineering Fair
www.sefi.org

Indianapolis
Indiana Science and Engineering Fair
http://www.sefi.org

Muncie
East Central Indiana Regional Science Fair
http://cms.bsu.edu/Academics/CollegesandDepartments/Biology/ResearchandCommunity/ECIScienceFair.aspx

South Bend
Northern Indiana Regional Science and Engineering Fair
http://science.nd.edu/nirsef

Valparaiso
Northwestern Indiana Science and Engineering Fair
http://www.valpo.edu/nwisef/

West Lafayette
Lafayette Regional Science and Engineering Fair
www.sefi.org/registration.html

KANSAS

Wichita
Kansas State Science and Engineering Fair
http://kssef.neosho.edu

KENTUCKY
Bowling Green
Southern Kentucky Regional Science Fair
http://skrsf.org/

Highland Heights
North Area Counties of Kentucky Exposition of Sciences
http://nackes.nku.edu
Lexington
Central Kentucky Regional Science and Engineering Fair
http://web.as.uky.edu/Biology/faculty/cooper/CKYscience/
Louisville
Louisville Regional Science Fair - Physical Science
http://www.LRSF.org
Louisville
Dupont Manual High School Regional Fair
http://www.dupontmanual.com
Richmond
Kentucky Science and Engineering Fair
http://www.kysef.eku.edu

LOUISIANA
Alexandria
Louisiana Region IV Science Fair

http://sciencefair.lsua.edu

Baton Rouge

Louisiana Region VII-Science and Engineering Fair

http://www.outreach.lsu.edu/lsef

Bossier City

Bossier Parish Community College Louisiana Region I Science and Engineering Fair

http://www.bpcc.edu/sciencefair.html

Hammond

Louisiana Region VIII Science Fair

http://www.selu.edu/acad_research/colleges/sci_tech/sci_fair/index.html

Houma

Terrebonne Parish Science Fair

Lafayette

Louisiana Region VI Science and Engineering Fair

Lake Charles

Louisiana Region V Science and Engineering Fair

http://www.lasciencefair.org

New Orleans

Greater New Orleans Science and Engineering Fair

http://gnosef.tulane.edu

Ruston

Louisiana Region II Science and Engineering Fair

https://sites.google.com/site/laregion2sciencefair/

St. James Parish

St. James Parish Science Fair

MASSACHUSETTS

Boston
Massachusetts Region VI Science Fair
http://www.scifair.com

Bridgewater
Massachusetts Region V Science Fair
Cambridge
Massachusetts State Science & Engineering Fair
http://www.scifair.com
Fall River
Massachusetts Region III Science Fair
North Adams
Massachusetts Region I Science Fair
http://www.mcla.edu/region1scifair
Somerville
Massachusetts Region IV Science Fair
http://shsscience.org/regionIV/
Worcester
Massachusetts Region II State Science Fair
http://www.wrsef.org

MARYLAND

Baltimore
Morgan State University Science-Mathematics-Engineering Fair
www.morgan.edu/centers/cemse
College Park

ScienceMontgomery - I
http://www.ScienceMONTGOMERY.org
Glen Burnie
Anne Arundel County Regional Science and Engineering Fair
http://www.aacps.org/science/secondscifair.asp
LaPlata
Charles County Science Fair
Largo
Prince George's Area Science Fair
Towson
Baltimore Science Fair
http://www.baltimoresciencefair.org
Walkersville
Frederick County Science and Engineering Fair
http://sciencesecondary.sites.fcps.org/countysciencefair2011

MICHIGAN

Ann Arbor
Southeast Michigan Science Fair
http://www.wccnet.edu/events/sciencefair/
Berrien Springs
Berrien County Regional Science Fair
Big Rapids
Mecosta-Osceola Science and Engineering Fair
http://www.moisd.org/sp/RSEF.htm
Detroit
Science and Engineering Fair of Metropolitan Detroit

http://www.sefmd.org

Flint

Flint Area Science Fair

http://www.flintsciencefair.org

Flint

Michigan Science and Engineering Fair

http://www.sciencefair.info

Kalamazoo

Southwest Michigan Science & Engineering Fair

Saginaw

Great Lakes Bay Region Science, Technology, and Engineering Fair

http://pub.spsd.net/communityeducation/SCSEF.htm

MINNESOTA

Bemidji

Northern Minnesota Regional Science Fair

Duluth

Northeast Minnesota Regional Science Fair

http://www.nemnscifair.org/

Mankato

South Central/SW Minnesota Regional Science and Engineering Fair

http://www.mnsu.edu/sciencefair

Minneapolis

Twin Cities Regional Science Fair

http://www.tcrsf.org

Minneapolis

St. Paul Science Fair

http://www.tcrsf.org

Minneapolis

Western Suburbs Science Fair

http://www.tcrsf.org

Minneapolis South

Minnesota Academy of Science State Fair

http://www.mnmas.org

Moorhead

Western Minnesota Regional Science Fair

http://www.mnstate.edu/scifair/

Rochester

Rochester Regional Science Fair

http://www.rochester.k12.mn.us

Saint Cloud

David F. Grether Central Minnesota Regional Science

http://web.stcloudstate.edu/jfmckenna/Regional%20Science%20Fair/sciencefair.html

Winona

Southeast Minnesota Regional Science Fair

http://www.winona.edu/sciencefair/

MISSOURI

Cape Girardeau

Southeast Missouri Regional Science Fair

http://www2.semo.edu/scifair

Fayette

Three Rivers Regional Science and Engineering Fair

Hillsboro
Mastodon Art/Science Regional Fair
http://www.MastodonFair.org
Jefferson City
Lincoln University Regional Science Fair
Joplin
Missouri Southern Regional Science Fair
http://www.mssu.edu/isef/
Kansas City
Greater Kansas City Science and Engineering Fair
http://www.sciencepioneers.org
Saint Joseph
Mid-America Regional Science and Engineering Fair
http://www.missouriwestern.edu/orgs/marsef/
Springfield
Ozarks Science and Engineering Fair
http://www.k12science.missouristate.edu
St. Louis
Academy of Science - Greater St. Louis Science Fair
http://sciencefairstl.org
St. Peters
Missouri Tri-County Regional Science and Engineering Fair

MISSISSIPPI
Biloxi
Mississippi Region VI Science and Engineering Fair
http://www.sciencefair.usm.edu

Booneville
Mississippi Region IV Science Fair
http://www.nemcc.edu

Greenville
Mississippi Region III Science and Engineering Fair
greenvilleweston.com/regioniii.html

Hattiesburg
University of Southern Miss. Region I Science and Engineering Fair
http://www.usm.edu/outreach/msef.html

Jackson
Mississippi Region II Science and Engineering Fair
http://www.jsums.edu/scifair

Mississippi State
Mississippi Region V Science and Engineering Fair
http://www.sciencefair.msstate.edu/

Oxford
Mississippi Region VII Science and Engineering Fair
http://www.outreach.olemiss.edu/youth/science_fair

Starkville
Mississippi Science and Engineering State Fair
http://www.usm.edu/outreach/msef.html

MONTANA

Billings
Billings Clinic Research Center Science Expo
http://billingsclinic.com/scienceexpo

Butte
Montana Tech Regional Science and Engineering Fair

http://mtechoutreach.org/sf/

Great Falls

Montana Region II Science and Engineering Fair

Havre

Hi-Line Regional Science and Engineering Fair - MSU-Northern

Missoula

Montana Science Fair

http://www.mtsciencefair.org/

NORTH CAROLINA

Charlotte

Charlotte-Mecklenburg Regional Science Fair

http://education.uncc.edu/cstem

Durham

North Carolina Central Region III Science Fair

http://nccrsf-3a.sfiab.com

Raleigh

North Carolina State Science Fair

http://www.ncsciencefair.org

NORTH DAKOTA

Fargo

Southeast North Dakota Regional Science and Engineering Fair

Grand Forks

Northeast North Dakota Regional Science and Engineering

Fair

Grand Forks

North Dakota State Science and Engineering Fair

Jamestown

Southeast Central North Dakota Science and Engineering Fair

Mandan

Southwest Central North Dakota Regional Science and Engineering Fair

Williston

Northwest North Dakota Regional Science Fair

NEBRASKA

Franklin

Central Nebraska Science and Engineering Fair

Nebraska City

Greater Nebraska Science and Engineering Fair

http://www.gnsef.org

NEW JERSEY

Jersey City

Hudson County Science Fair

Lawrenceville

Mercer Science and Engineering Fair

http://mercersec.org

New Brunswick

North Jersey Regional Science Fair

http://njrsf.org/

NEW MEXICO

Albuquerque

National American Indian Science and Engineering Fair

http://www.aises.org

Albuquerque

Central New Mexico Regional Science and Engineering Challenge

http://stemed.unm.edu

Farmington

San Juan New Mexico Regional Science and Engineering Fair

Grants

Four Corners Regional Science and Engineering Fair

Las Cruces

Southwestern New Mexico Regional Science and Engineering Fair

http://scifair.nmsu.edu/

Las Vegas

Northeastern New Mexico Regional Science and Engineering Fair

http://www.nmhu.edu/communityeducators/educators/sciencefair

Portales

Southeastern New Mexico Regional Science and Engineering Fair

Socorro

New Mexico Science and Engineering Fair - Life Sciences

http://www.nmt.edu/~science/fair

NEVADA
Elko
Elko County Science Fair
Las Vegas
Southern Nevada Regional Science and Engineering Fair
Reno
Western Nevada Regional Science and Engineering Fair
http://www.nevadasciencefair.net

NEW YORK
Long Island
Long Island Science and Engineering Fair
http://www.lisef.org
New York City
New York City Science and Engineering Fair
http://collegenow.cuny.edu/sciencefair
Poughkeepsie
Dutchess County Regional Science Fair
http://www.dcsciencefair.org
Queens
New York State Science and Engineering Fair
http://www.nyssef.org
Rochester
Central Western Section-Science Teachers Association of New York State Science Congress

http://www.ggw.org/~cws

Syracuse

Greater Syracuse Scholastic Science Fair

http://www.most.org

Syracuse

Dr. Nelson Ying Tri Region Science and Engineering Fair

http://www.YingTRSEF.org

Troy

Greater Capital Region Science and Engineering Fair

http://www.gcrsef.org

7 Joshua Street, Saratoga Springs, 12866

Utica

Utica College Regional Science Fair

http://utica.edu/sciencefair

Westchester

Progenics Westchester Science and Engineering Fair

http://www.wesef.org

OHIO

Alliance

Ohio Region XIII Science and Engineering Fair

http://raider.mountunion.edu/Organizations/scienceday/

Archbold

Northwest Ohio Science and Engineering Fair

Athens

Southeastern Ohio Regional Science and Engineering Fair

http://www.ohiou.edu/scifair/

Cleveland

Northeastern Ohio Science and Engineering Fair

http://neosef.org

Columbus

Buckeye Science and Engineering Fair

http://www.ohiosci.org

Dayton

Montgomery County Science and Engineering Fair

http://www.montgomerycountyscienceday.org

Marion

Marion Area Science and Engineering Fair

http://academic.marion.ohio-state.edu/sciencefair

Shaker Heights

Hathaway Brown Upper School Fair

Upper Arlington

Central Ohio Regional Science and Engineering Fair

http://www.corsef.org

Wilberforce

Miami Valley Science and Engineering Fair

http://www.centralstate.edu/academics/sciencefair

OKLAHOMA

Ada

East Central Oklahoma Regional Science and Engineering Fair

Ada

Oklahoma State Science and Engineering Fair

http://ossef.ecok.edu

Alva

Northwestern Oklahoma State University Regional Science Fair
http://www.nwosu.edu/science/fair

Bartlesville
Bartlesville District Science Fair
http://www.BartlesvilleDSF.org

Edmond
Central Oklahoma Regional Science and Engineering Fair
http://cms.uco.edu/ScienceFair/

Miami
Northeastern Oklahoma A&M Science and Engineering Fair
http://www.neo.edu/sef

Muskogee
Muskogee Regional Science and Engineering Fair

Oklahoma City
Oklahoma City Regional Science and Engineering Fair
http://www.okcu.edu/biology/fair.aspx

Wilburton
Eastern Oklahoma Regional Science and Engineering Fair
http://www.eosc.edu/academic/science_div.html

OREGON

Albany
Central Western Oregon Science Expo
http://nwse.org/cwose

Bend
High Desert Science Expo

Gresham

Gresham-Barlow Science Expo
http://nwse.org/

Hillsboro
Beaverton-Hillsboro Science Expo
http://bhse.org

Portland
Portland Public Schools Science Expo
http://www.nwse.org

Portland
Aardvark Science Exposition
http://www.oes.edu/us/departments/science-expo.html

Portland
Intel Northwest Science Expo
http://www.nwse.org

West Linn
CREST-Jane Goodall Science Symposium
http://www.crest.wlwv.k12.or.us/

PENNSYLVANIA

Harrisburg
Capital Area Science and Engineering Fair
http://www.whitakercenter.org/CASEF.html

Lancaster
North Museum Science and Engineering Fair
http://www.northmuseum.org/ScienceFair/tabid/221/Default.aspx

Philadelphia

Delaware Valley Science Fairs
http://www.dvsf.org
Pittsburgh
Pittsburgh Regional Science and Engineering Fair
http://www.pittsburghsciencefair.org
Reading
Reading and Berks Science and Engineering Fair
http://www.rbsef.org
York
York County Science and Engineering Fair
http://ycsef.org/YCSEF/Home.html

RHOSE ISLAND

Warwick
Amgen Rhode Island Science and Engineering Fair
http://risef.org

SOUTH CAROLINA

Aiken
Central Savannah River Area Science and Engineering Fair
http://educationoutreach.srs.gov/srrsef.htm

Charleston
Low Country Science Fair
http://www.cofc.edu/lowcountryhall/lsf.htm
Columbia
USC Central South Carolina Region II Science and Engineering Fair

http://www.cas.sc.edu/cse/SF/index.htm

Florence

Sand Hills Regional Science Fair

Hilton Head Island

Sea Island Regional Science Fair

Spartanburg

Piedmont South Carolina Region III Science Fair

SOUTH DAKOTA

Aberdeen

Northern South Dakota Science and Math Fair

Brookings

Eastern South Dakota Science and Engineering Fair

http://www.sdstate.edu/sciencefair

Mitchell

South Central South Dakota Science and Engineering Fair

http://www.dwusciencefair.com

Rapid City

High Plains Regional Science and Engineering Fair

http://www.hpcnet.org/science

TENNESSEE

Chattanooga

Chattanooga Regional Science and Engineering Fair

http://www.chattanoogasciencefair.org

Clarksville

Middle Tennessee Science and Engineering Fair

http://www.apsu.edu/mtsef

Cookeville

Cumberland Plateau Regional Science and Engineering Fair

www.tntech.edu/stem

Jackson

West Tennessee Regional Science and Engineering Fair

http://www.uu.edu/events/wtrsef/

Knoxville

Southern Appalachian Science and Engineering Fair

http://www.sasef.com

Memphis

Memphis-Shelby County Science and Engineering Fair - North Region

http://mscsef.mecca.org/

TEXAS

Austin

Austin Energy Regional Science Festival

http://www.sciencefest.austinenergy.com

Brownsville

Rio Grande Valley Regional Science and Engineering Fair

http://www.rgvrsef.com

Dallas

Beal Bank Dallas Regional Science and Engineering Fair

http://DallasScienceFair.org

El Paso

Sun Country Science Fair

http://www.hsaelpaso.org/sciencefair/index.html

Fort Worth
Fort Worth Regional Science and Engineering Fair
http://www.fwrsef.org
Houston
Science Engineering Fair of Houston
www.sefhouston.org
Kilgore
East Texas Regional Science Fair
http://www.kilgore.edu/chemistry_fair.asp
Laredo
United Independent School District Regional Science Fair
http://www.uisd.net
Laredo
Laredo Independent School District Science Fair
Lubbock
South Plains Regional Science and Engineering Fair
Odessa
Permian Basin Regional Science Fair
http://www.utpb.edu/scifair/
San Angelo
District XI Texas Science Fair
http://www.angelo.edu/org/science_competition/
San Antonio
Alamo Regional Science and Engineering Fair
http://www.arase.org
San Antonio
ExxonMobil Texas Science and Engineering Fair
http://emtsef.utsa.edu/

Waco
Central Texas Science and Engineering Fair
http://www.ctsef.org

UTAH
Cedar City
Southern Utah Science and Engineering Fair
http://suu.edu/sci/fair/
Layton
North Davis Area Science and Engineering Fair
http://www.davis.k12.ut.us/district/curric/science/links.html
Ogden
Weber Area Science and Engineering Fair
http://curriculum.weber.k12.ut.us/scifair/index.htm
Ogden
Harold W. & Helen M. Ritchey Science and Engineering Fair of Utah
http://www.ritcheysciencefair.org/
Provo
Central Utah Science and Engineering Fair
http://cusef.byu.edu
Salt Lake City
Salt Lake Valley Science and Engineering Fair
http://slvsef.org

VIRGINIA
Arlington
Northern Virginia Science and Engineering Fair
http://www2.apsva.us/1540108217759100/blank/browse.asp?

A=383&BMDRN=2000&BCOB=0&C=54442

Charlottesville

Virginia Piedmont Regional Science Fair

http://www.vprsf.org

Fairfax

Fairfax County Regional Science and Engineering Fair

http://www.fcps.edu/dis/sciengfair/

Harrisonburg

Shenandoah Valley Regional Science Fair

http://csm.jmu.edu/svrsf

Lynchburg

Central Virginia Regional Science Fair

http://www.cvgs.k12.va.us/scifair

Manassas

Prince William-Manassas Regional Science Fair

http://pwcs.edu

Norfolk

Tidewater Science Fair

http://www.cs.odu.edu/~scifair/RegionalFair/index.html

Norfolk

Virginia State Science and Engineering Fair

www.vssef.odu.edu

Purcellville

Loudoun County Science and Engineering Fair

http://www.lcps.org/rsef

Radford

Blue Ridge Highlands Regional Science Fair

Richmond (03/19/2011 - 03/19/2011) Metro Richmond

Science Fair
http://msinnovation.info/scifair/index.htm
Roanoke
Western Virginia Regional Science Fair

VERMONT

Northfield
Vermont State Science and Mathematics Fair
http://vssmf.pbworks.com/FrontPage
Bellevue
Central Sound Regional Science & Engineering Fair
http://scidiv.bellevuecollege.edu/SAMI/scifair
Bremerton
Washington State Science and Engineering Fair
http://www.wssef.org/
Kennewick
Mid-Columbia Regional Science and Engineering Fair
http://www.mcsf.net
Tacoma
South Sound Regional Science Fair
http://www.plu.edu/~scifair

WISCONSIN

Glendale
Nicolet Science and Engineering Fair
Madison
Capital Science and Engineering Fair
http://capitalsciencefair.org

Milwaukee
University School of Milwaukee - Science Fair
http://www.usmk12.org
Milwaukee
Badger State Science and Engineering Fair
http://www.BSSEF.org
Sheboygan
Lakeland Science and Engineering Fair
http://lcsef.lakeland.edu/

WEST VIRGINIA

Fairmont
West Virginia State Science and Engineering Fair
http://www.fairmontstate.edu/academics/CollegeofSciTech/WVSSEF.asp
Keyser
West Virginia Eastern Panhandle Regional High School Science Fair
Montgomery
Central and Southern West Va. Regional Science and Engineering Fair
http://sciencefair.wvutech.edu
West Liberty
West Liberty State University Regional Science and Engineering Fair
http://www.westliberty.edu

WYOMING

Greybull

Northern Wyoming District Science Fair

Laramie

Wyoming State Science Fair

www.uwyo.edu/sciencefair

Made in the USA
Lexington, KY
07 April 2014